Springer Theses

Recognizing Outstanding Ph.D. Research

For further volumes:
http://www.springer.com/series/8790

Aims and Scope

The series "Springer Theses" brings together a selection of the very best Ph.D. theses from around the world and across the physical sciences. Nominated and endorsed by two recognized specialists, each published volume has been selected for its scientific excellence and the high impact of its contents for the pertinent field of research. For greater accessibility to non-specialists, the published versions include an extended introduction, as well as a foreword by the student's supervisor explaining the special relevance of the work for the field. As a whole, the series will provide a valuable resource both for newcomers to the research fields described, and for other scientists seeking detailed background information on special questions. Finally, it provides an accredited documentation of the valuable contributions made by today's younger generation of scientists.

Theses are accepted into the series by invited nomination only and must fulfill all of the following criteria

- They must be written in good English.
- The topic should fall within the confines of Chemistry, Physics, Earth Sciences, Engineering and related interdisciplinary fields such as Materials, Nanoscience, Chemical Engineering, Complex Systems and Biophysics.
- The work reported in the thesis must represent a significant scientific advance.
- If the thesis includes previously published material, permission to reproduce this must be gained from the respective copyright holder.
- They must have been examined and passed during the 12 months prior to nomination.
- Each thesis should include a foreword by the supervisor outlining the significance of its content.
- The theses should have a clearly defined structure including an introduction accessible to scientists not expert in that particular field.

Wen-Te Liao

Coherent Control of Nuclei and X-Rays

Doctoral Thesis accepted by
the University of Heidelberg, Germany

Author
Dr. Wen-Te Liao
Theory Division
Max Planck Institute for Nuclear Physics
Heidelberg
Germany

Supervisors
Honorarprof. Dr. Christoph H. Keitel
Dr. Adriana Pálffy
Theory Division
Max Planck Institute for Nuclear Physics
Heidelberg
Germany

ISSN 2190-5053 ISSN 2190-5061 (electronic)
ISBN 978-3-319-35024-0 ISBN 978-3-319-02120-1 (eBook)
DOI 10.1007/978-3-319-02120-1
Springer Cham Heidelberg New York Dordrecht London

Printed on acid-free paper

Springer is part of Springer Science+Business Media (www.springer.com)

To my grandmother Yu-Fong Liao

Supervisor's Foreword

More than 50 years ago, the invention of the laser came to revolutionize atomic physics in an unprecedented and rather unexpected manner. Since 1960, optical coherent fields have been used to literally control atoms and atomic transitions, leading not only to a better understanding of atomic and molecular dynamics, but also to a vast number of applications. While experimental developments have allowed a steady rise of available laser intensities in the last few decades, laser frequencies, until recently, remained restricted to the optical. The commissioning of the first X-ray free electron laser in 2009 opened new opportunities for the study of phenomena in the interaction of high-frequency lasers with matter.

With coherent X-rays available, a tool may emerge for the control of nuclear transitions. Nuclei with low-lying excited states at energies of several tens of keV are natural candidates for the interaction with coherent light generated by X-ray free electron lasers. An important novel aspect here, compared to atoms, is the interplay that occurs between the strong and electromagnetic forces, which in turn means that the induced coherent effects may be small and difficult to observe. In addition, a number of purely nuclear incentives support the quest for coherent control of nuclei. Among these, the most prominent are the design of a gamma-ray laser based on nuclear transitions, the release on demand of energy stored in nuclear isomers and the development of a nuclear frequency standard. As a step towards these goals, the present thesis investigates theoretically the exciting prospects for the interaction of coherent light with nuclei.

Coherent control of atoms has already been mastered with quantum optics techniques. It is therefore natural that one may first look into the possibility of applying acknowledged quantum optics schemes to nuclear systems. This thesis thus addresses the possibility of manipulating the population of nuclear states using strong coherent X-ray fields and, in particular, of controlling the pumping and decay of long-lived excited levels—known as isomers—that store nuclear energy over long periods of time. If experimentally successful, this idea may open up an avenue to new nuclear energy storage solutions that involve neither fission nor fusion. Apart from nuclear batteries, specific isomers offer a playground for establishing a new level of precision for a frequency standard based on nuclear transitions. In this case, too, tricks originally developed in atomic quantum optics turn out to be very useful for improving the nuclear level measurement precision up the accuracy required for a ultra-precise nuclear clock.

While in the future very strong X-ray fields may be used to control nuclei, one may also turn the problem around to see how X-ray light might be controlled by nuclei. Instead of merely pumping a nuclear transition by extremely strong lasers, one envisages the situation of a weak excitation—a single nucleus, in fact—that stores a single X-ray photon. Presented here is a novel scenario to store, release and control the properties of such a photon by using a "nuclear cage". This scenario creates an important missing link between nuclei and photonic devices and photon qubits. Will nuclei and X-rays in the future provide the physical basis for our computers, evading the bottleneck of the diffraction limit of the optical frequencies? Are we now on the verge of a revolution in nuclear physics, as atomic physics was back in the days of Ted Maiman's first laser? And what other surprises can be expected from coherent control of nuclei, so far little more than terra incognita? A promising start awaits the reader in the pages of this work.

Heidelberg, July 2013

Christoph H. Keitel
Adriana Pálffy

Abstract

The possibility of mutual control of nuclei and photons offered by the emerging field of nuclear quantum optics is theoretically investigated in three different applications. This extension of quantum optics towards nuclei is motivated by modern X-ray Free Electron Lasers (XFEL) which open the possibility to coherently control nuclear states. As a first application we investigate the coherent population transfer between nuclear states in a three-level system driven by an XFEL. Such a level scheme is relevant for the triggering of isomers and might play a role for future energy storage solutions. The other way around, nuclei offer a platform to control single X-ray photons, which can be focused on spots essentially smaller than a single atom and used in future photonic circuits. The second part of this thesis puts forward a nuclear forward scattering setup that allows coherent control of a single X-ray photon using ^{57}Fe nuclei. Finally, we show that nuclear quantum optics provides a significant improvement for detection in nuclear physics. The low-lying isomeric transition of ^{229}Th can be addressed by VUV lasers and provides a potential next generation frequency standard. A main impediment is the large uncertainty of the nuclear transition frequency. Using an electromagnetically modified nuclear forward scattering setup, we show that coherence effects can reduce this uncertainty down to an unprecedented level.

Acknowledgments

It is with great gratitude that I acknowledge the support of my Supervisor, Honorarprof. Christoph H. Keitel. He is an inspiring Teacher, and always took the time to listen to my opinions and gently pointed out potential risks. His passion for physics encouraged me—I will always remember the fervor in his eyes when he first explained to me the idea of nuclear batteries. Additionally, at each meeting Prof. Keitel's humor always lightened the exciting discussions of physics, and his words animated me to greater effort. I would like to thank him for accepting me as a Ph.D. student in the Theory division at the Max Planck Institute for Nuclear Physics and for the financial support since April 2010.

I am sincerely grateful to my Supervisor, Dr. Adriana Pálffy, for her great help throughout my Ph.D. career in Germany since April 2010. She was always very patient with my immature opinions, and untiringly taught me. Adriana's supervision was excellent, and the collaboration with her was enjoyable. Furthermore, I cannot find words to express my gratitude to Adriana for the guidance she showed me throughout my dissertation writing, especially during the period of her pregnancy. May God bless her and her baby.

I would like to thank my Supervisor, Prof. Selim Jochim for giving me many useful suggestions and comments.

I would like to thank Dr. Andrey Surzhykov for reading my thesis and writing the second referee report.

I would like to thank the further members of my examination committee Prof. Markus Oberthaler and Prof. Klaus Pfeilsticker for their interest in my research topic.

I would like to thank our Secretaries Sibel Babacan and Ludmila Hollmach for their help in bureaucracy aspects. My special thanks go to Sibel Babacan for her great help since the interview of International Max Planck Research School for Quantum Dynamics in December 2009.

I am indebted to my many Proof Readers Dr. Felix Mackenroth, Kilian P. Heeg, Jonas Gunst, Dr. Sven Ahrens, Stefano Cavaletto and Andreas Reichegger, who carefully read the early versions of my thesis and gave me their useful comments and hints. My special thanks go to Dr. Felix Mackenroth for his great help in good English language proficiency, and also to Kilian P. Heeg, Martin Gärttner and Dr. Felix Mackenroth for their German language assistance.

I wish to thank Dr. Jörg Evers for many suggestions and interesting discussions in quantum optics.

I would like to thank Dr. Héctor Mauricio Castañeda Cortés, Dr. Benjamin J. Galow, Dr. Felix Mackenroth, Dr. Sandra I. Schmid and Dr. Sven Ahrens for their advice considering the final exam.

I wish to thank Dr. Ralf Röhlsberger for his many suggestions to the ^{57}Fe project. Especially, I am indebted to his fruitful knowledge of nuclear scatterings and his kind answers to my many questions.

I wish to thank Dr. Sumanta Das for the good time of our cooperation in the ^{229}Th project.

I wish to thank Dr. Thorsten Peters, Dr. Thomas Herrmannsdörfer and Prof. Yutaka Nomura for their scientific advice.

I wish to thank many members of the thorium group at the Institute of Atomic and Subatomic Physics in Vienna Dr. Thorsten Schumm, Dr. Georgy Kazakov and Matthias Schreitl for the very fruitful discussions on theoretical as well as experimental challenges. My special thanks go to Dr. Georgy Kazakov for his suggestions and selfless help to the ^{229}Th project.

I wish to thank Prof. Girish S. Agarwal for the inspiring discussions in quantum optics during his visit at the Theory division at the Max Planck Institute for Nuclear Physics in May 2012.

I wish to thank Prof. Markus Aspelmeyer for inviting me to visit his group in Vienna and the exciting discussions in September 2012. My special thanks go to his Secretary, Alexandra Seiringer, for her help in bureaucracy aspects, and also to Dr. Nikolai Kiesel for his great help and the fruitful discussions.

I wish to thank Prof. Ite A. Yu and Prof. Shih-Chuan Gou for writing the referee letters to support my application on the Ph.D. scholarship in the International Max Planck Research School for Quantum Dynamics in October 2009.

I wish to thank Yu-Jen Lin for kindly providing the very beautiful rainbow photo and the great time together.

I would like to thank Dr. Angela Lahee for her interest in my thesis. My special thanks go to Mr. Anand Jayaprakash for his help in bureaucracy aspects.

My sincere thanks go to my beloved Yu-Yen Chang for her personal support, the wonderful time together, and her patience throughout my dissertation writing.

Last but not least, I am sincerely grateful to my grandmother Yu-Fong Liao, my father Jhen-Cheng Liao, my mother A-Jin Yang, my young sister Wen-Hua Liao and all my family. With your understand and support, I am able to concentrate on my work and be the one I am.

Contents

Related Publications

Within the framework of this thesis, the following articles were published in refereed journals:

- Wen-Te Liao, Adriana Pálffy and Christoph H. Keitel,
  Nuclear Coherent Population Transfer with X-Ray Laser Pulses, Phys. Lett. B **705**, 134 (2011), arXiv:1011.4423.
- Wen-Te Liao, Adriana Pálffy and Christoph H. Keitel,
  Coherent Storage and Phase Modulation of Single Hard-X-Ray Photons Using Nuclear Excitons, Phys. Rev. Lett **109**, 197403 (2012), arXiv:1205.5503. (Featured by D. Lindley, "Focus: Storing an X-ray Photon", Physics 5, 125 (2012); also see "Gespeicherte Röntgenphotonen" the press release of the Max Planck Institute for Nuclear Physics, http://www.mpi-hd.mpg.de/mpi/en/news/meldung/detail/gespeicherte-roentgenphotonen/)
- Wen-Te Liao, Sumanta Das, Christoph H. Keitel and Adriana Pálffy,
  Coherence Enhanced Optical Determination of the ^{229}Th Isomeric Transition, Phys. Rev. Lett **109**, 262502 (2012), arXiv:1210.3611.
- Wen-Te Liao, Adriana Pálffy and Christoph H. Keitel,
  Three-Beam Setup for Coherently Controlling Nuclear-State Population, Phys. Rev. C **87**, 054609 (2013), arXiv:1302.3063. (Editor's suggestion)

Articles in preparation:

- Wen-Te Liao, Adriana Pálffy and Christoph H. Keitel, *Continuous Phase Modulation of Single Hard X-Ray Photon Wavepackets.*
- Wen-Te Liao, Adriana Pálffy and Christoph H. Keitel, *Optical Determination of the ^{229}Th Isomeric Transition Using Magnetically Induced Quantum beats.*

Chapter 1
Introduction

"Let there be light" [1] and there was the vivid world. A rainbow in the sky, a glance of Mona Lisa's smile and the New Year's Eve fireworks at the Taipei 101 are just few gorgeous views mediated by photons. Additionally, the cosmic microwave background radiation discloses the secrets of the early Universe [2] and x-rays scattered off a crystal reveal its structure [3] to name a few examples. Mankind may have realized the importance of electromagnetic waves via appreciating natural phenomena, and thus started to investigate the interaction between light and matter. This development is paramount in the history of science and successfully teaches us how to get control over both matter and photons. For example, Isaac Newton demonstrated that a prism could decompose white light into a spectrum of colors, and a modern extension of this spectral decomposition led to nowadays ultrashort and strong-field laser technology [4]. The photons from a star reveal the Hanbury Brown and Twiss effect and show the nature of correlation between individual photons [5]. In the neighbouring field of biology, the fatal beauty of the shiny jellyfish Aequorea victoria may have attracted scientists to discover and to isolate the green fluorescent protein (GFP) from it [6]. After many investigations, the emission off GFP became an indispensable tool for studying molecular biology as it displays the operations of living cells [7]. Similar impact on nuclear physics is also given by studying the photons that interact with nuclei.

In general, nuclear transition energies are above 10 keV, and the momentum of the corresponding γ-ray photon that drives the transition is at least 4 orders of magnitude higher than that of optical photons. Due to the very large photon momentum, the absorption or the emission of a γ-ray photon by a nucleus usually reveals an obvious line shift due to the conservation of momentum in nuclear spectroscopy. In 1958, R. Mössbauer turned to this problem and observed the resonant recoil-free absorption of γ-rays by atomic nuclei bound in a solid sample [8, 9]. For the first time, scientists could get rid of recoil by having it absorbed by the solid-state lattice as a whole and actively observe the undisturbed nuclear transitions. However, the technique of Mössbauer spectroscopy is limited by the need for a suitable nuclear transition and the corresponding photon source which typically consists of radioactive parent

W.-T. Liao, *Coherent Control of Nuclei and X-Rays*, Springer Theses,
DOI: 10.1007/978-3-319-02120-1_1,

Fig. 1.1 A rainbow in the sky over Daguan mountain (La La mountain), Taiwan. Photo provided by Yu-Jen Lin

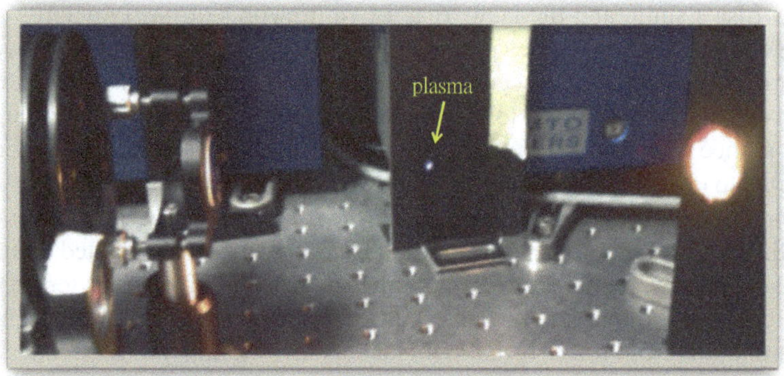

Fig. 1.2 Plasma produced by a modern strong-field laser. Laser pulses are focused on the tiny bright spot in the air, where the laser power is high enough to ionize it. Photo of the setup in Thomas Pfeifer's group at the Max Planck Institute for Nuclear Physics in Heidelberg, Germany

nuclei decaying to excited states of the studied isotopes. Because of this limitation, Mössbauer spectroscopy is only applicable to particular nuclear transitions, e.g., the lowest nuclear transition of ^{57}Fe at 14.4 keV. Until in the 1970s the second-generation synchrotron light sources started operation and provided much brighter broadband photon beams of such short wavelengths. Eventually inspired by this technological advance in 1974, S. Ruby suggested to replace the usual radioactive γ-ray sources with synchrotron radiation [10–12], such that more species of nuclei could be studied [13]. This idea led to the development of nuclear forward scattering (NFS) which has raised to a mature level of applications as it provides scientists with the possibility to explore nuclear condensed matter physics with synchrotron radiation [13].

Driving nuclear transitions with light can have also other practical motivations than the original experiments by R. Mössbauer. This is related to the existence of

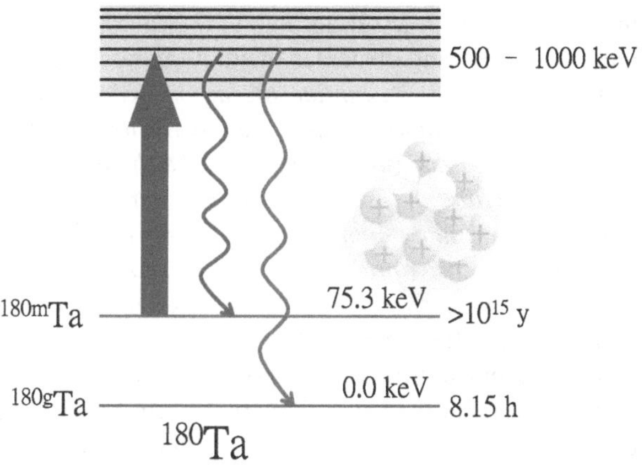

Fig. 1.3 Isomer triggering of the ^{180m}Ta state. The green vertical wide arrow shows the incident incoherent triggering radiation and the purple (*green*) wiggled arrow illustrates the depopulation from some excited triggering states to the ground (isomer) state. ^{180m}Ta and ^{180g}Ta denote the isomer and the ground state, respectively

long-lived excited nuclear states, known as nuclear isomers [14], that can store large amounts of energy over long periods of time [14, 15]. If efficiently controlled, such states could be used as transportable nuclear batteries, by storing or releasing on demand the excitation energy. The direct decay of an isomer to the ground state is however strongly suppressed. Releasing the stored energy can therefore rather be achieved by driving a nuclear transition to an excited triggering level above the isomer, a process known as isomer triggering (IT) or isomer depletion [15]. This excited level can then decay unhindered to the ground state, thus releasing the stored energy. A schematical picture of this process is presented in Fig. 1.3. The red line and blue line illustrate the isomer and the ground state of a nucleus, respectively. An incoherent radiation illuminates a nuclear sample to pump the isomeric population to some excited triggering states, and then the excited population will spontaneously decay to the nuclear ground level. This IT process releases the stored energy as photons emitted in the spontaneous decay. By using this method, as showed in Fig. 1.3, a demonstration of IT was performed in 1999 by shinning incoherent 6 MeV bremsstrahlung radiation on a ^{180}Ta sample [16]. However, this kind of IT relies on incoherent processes, e.g., pumping by incoherent radiation and depopulation through spontaneous decay, and the efficiency is rather low.

From the study of NFS and incoherent IT, one realizes that using an incoherent electromagnetic wave like synchrotron radiation to control or excite nuclei is rather inefficient. In contrast to that, coherent control as known in quantum optics with visible photons would provide high efficiency. Just like a group of dancers dance with the music, in physics, a coherent laser is the "music" which makes a large number of charged particles, e.g., nuclei simultaneously oscillate together. This kind of conducted behaviour is labeled as coherent control, and scientists utilize this method to manipulate atomic and molecular systems with optical lasers. However, due to the lack of coherent γ-ray sources [17], such high efficient control could not be achieved so far in any nuclear system. Technological limitations were the main reason to consider many research directions with nuclei as "mission impossible" in the last century. J. M. Blatt and V. F. Weisskopf in 1979 summarized some of the then predominating obstacles of studying nuclei and the radiation emitted by nuclei in their book [18] *"...unlike the atomic case, the wavelengths of interest are so short that they cannot be measured by the usual optical devices. The rather indirect methods which must be employed make the energy determination quite inaccurate ...and the available resolution low. Furthermore in most cases the radiation process is only one of many competing processes (such as particle emission or sometimes beta-decay) and its probability is correspondingly low. For these reasons the study of gamma-rays from nuclei has remained on a rather rudimentary level ..."*.

However, after many efforts, some key obstacles have been overcome, e.g., x-ray coherent sources have become available. In 2009, the first X-ray Free Electron Laser (XFEL) at the Linac Coherent Light Source in Stanford (USA) fired up [19], delivering very bright and partially coherent hard x-ray flashes. Moreover, several upgrades of the XFEL are already commissioned and produce fully coherent x-ray laser [20]. Apart from the new light sources like the XFEL [21–26], detectors with a better energy resolution [27] and the improved optics elements for hard x-rays [28–31] contribute to the breakthrough. The totality of all these technology realizations may lead us into an entirely new era of studying the interaction between nuclei and radiation, i.e., nuclear quantum optics.

The coherent manipulation of nuclei by XFEL photons or in turn of x-ray photons using nuclear transitions open a vast land of unexplored opportunities with possible applications related to nuclear batteries or even quantum information and computing. Motivated by the eager anticipation, this dissertation is devoted to a theoretical demonstration of three new proposals for the emerging field of nuclear quantum optics and the joy of the physics behind. A first topic addressed is related to coherent nuclear control with x-ray lasers. In particular, we study the possibility of nuclear coherent population transfer (NCPT) with x-ray pulses. This is completely revolutionary and provides an early stage for investigating the coherent isomer triggering or nuclear battery, a clean and safe solution for the energy storage. Some early key proposals of the XFEL-based nuclear coherent control were, for example, the nuclear Rabi oscillation [32] or laser-driven direct quantum control of nuclear excitations [33]. However, a coherent IT or NCPT between nuclear ground state and an isomer state with XFEL pulses was never addressed before. For the first time, this topic is discussed in Chap. 3 of this thesis entitled **Nuclear Coherent Population Transfer with X-Ray Laser Pulses**.

A clean nuclear energy storage may ease the energy crisis ahead, and laser induced isomer triggering or population, as mentioned above, is one of the most promising solutions. The advance of modern XFEL provides completely new opportunities to control nuclear states with very bright coherent hard x-ray beams from XFEL facilities. Anticipating this coming revolution, in this chapter the well-known stimulated Raman adiabatic passage and the two π-pulses method are applied for the first time to control a nuclear quantum state with two Lorentz-boosted XFEL pulses [34]. We find that efficient NCPT is already achievable with laser intensities of 10^{18}–10^{19} W/cm^2 which will be available at facilities in commissioning at present. In addition to the consideration of large facilities, the possibility of using tabletop setups for NCPT is also discussed, in conjunction with the application of IT and isomer population for nuclear batteries.

If coherent light can be used to control nuclear transitions, this is also valid the other way around. Nuclei can be used to control light, especially on the level of single photons. As a second main topic of this thesis, we address the coherent control of single hard x-ray photons using nuclei. In the past, the main impediment for this type of study were the poor x-ray optics elements, making a high energy photon rather a tool instead of an easily manipulated subject. Due to significant technological advances, modern optics elements were significantly improved for low energy γ-rays (hard x-rays), e.g., the development of diamond mirrors with more than 99 % reflectivity [28, 35], hard x-ray waveguides [29, 30] and the Fabry-Pérot resonator [31, 36, 37]. Hence the road for studying single hard x-ray photons is paved. Hard x-ray photons may interact with inner shell electrons of an atom [38, 39], highly charged ions [40, 41] or alternatively with nuclei [42, 43]. Some advantages of using nuclei are:

1. A nuclear excitation created by the absorption of hard x-rays is reversible, whereas an atom may be ionized by the x-ray photon absorption [39].
2. Nuclei can be embedded in a solid-state sample [42–45] such that one does not need a very complicated ion trap [40, 41] or a gas cell [38, 39] to contain them.
3. The absorption and the emission of hard x-rays by a nucleus bound in a solid sample can be recoilless, such that the whole sample experiences the same nuclear transition frequency.
4. The number of nuclei bound in a crystal [42–45] interacting with photons can be many orders of magnitude greater than the number of trapped ions in a gas cell [38, 39, 41] or an ion trap.
5. The lifetime of a nuclear solid-state sample is essentially infinite compared to that of an ion gas sample. Therefore a nuclear solid-state sample can be used as a tiny optics element.
6. Doped dense nuclei in a crystal are less sensitive to their environments, whereas trapped dense ionized atoms interact with each other.

The drawback of studying hard x-rays from nuclei are the small cross Sections [46] which, however, can be overcame by the high abundance of nuclei in a crystal. Recent key studies using nuclei for such experiments were the incoherent storage of hard x-ray single photons [42], producing keV single photon entanglement [47], controlling the absorption of hard x-rays via nuclear level anticrossing in a $FeCO_3$ crystal [48], measurement of collective Lamb shift [49], the implementation of electromagnetically induced transparency (EIT) with resonant nuclei [50] in x-ray cavities, or coherent optical control of Mössbauer spectra [51].

Motivated by this, in Chap. 4 entitled **Coherent Storage and Phase Modulation of Single Hard X-Ray Photons** we study the possibility to control a single x-ray photon wavepacket with iron nuclei. In a way, this reminds of the science fiction scenario in B. Shaw's story *Light of Other Days*, which introduces the fascinating concept of "slow glass". The refractive index of slow glass is so high that light needs one year to slowly penetrate it, such that one can only see the past through it. Indeed, one can manipulate the behavior of photons by controlling the refractive index of matter. For example, if one can dynamically change the refractive index of a "slow glass", the image of the past traveling through the glass can be accelerated towards the viewer by lowering the index, or the glass suddenly becomes dark for a very high index. For optical photons, control of the refractive index and slow light with a group velocity of about 17 m/s have been achieved in experiments [52].

This chapter aims at demonstrating the above scenario for low-energy γ-rays and proposing an iron (^{57}Fe) nuclear cage for single hard x-ray photons. With our proposal, one essentially could store a single hard x-ray photon in a ^{57}Fe solid-state sample, e.g., stainless steel, and retrieve it later without losing its original properties. Also, our proposal is based on the well developed NFS technique, and its implementation is therefore not far-fetched. Since the spot size of a focused hard x-ray can essentially be smaller than the size of a single atom, one can think about treating a single atomic nucleus as a memory. Because of this possibility, the coherent storage of single hard x-ray photons may have an impact on the architecture of future quantum memories. Furthermore, we demonstrate the possibility of manipulating the phase of the stored photon, through which the released "past" x-ray photons from "frozen stainless steel" can be customized. In the future, the shops may stock the "X-Rays of Other Days" stored in a nuclear memory.

For atomic systems, coherent control has been very successful leading apart from the development of fields such as quantum information also to a number of every-day life applications, for instance the Global Positioning System (GPS) [53]. If iron nuclei might one day serve as a very compact quantum memory, another nuclear transition might also provide the new generation of high precision clocks. So far, the global frequency standard is based on Cs atomic fountain clocks which have reached an uncertainty level of 4.5×10^{-16} [54]. To establish a more precise definition for time and length, one has to refer to a higher energy quantum oscillator and radiation with a shorter wavelength. Thus, an atomic nucleus and electromagnetic waves in the spectrum from VUV to hard x-rays are good candidates to construct such high precision systems. In addition to the recently developed XUV [55, 56] and the proposed x-ray [57] frequency combs, the spotlighted ideas in this direction are a ^{229}Th "nuclear clock" [58, 59] and a combined optical-x-ray Fabry-Pérot interferometer [37, 60]. The ^{229}Th nucleus has a very narrow isomeric state at about 7.8 eV which is already accessible for VUV lasers which recommends it for a new generation of high-precision clocks based on nuclei. However, at present the accuracy of the experimentally determined nuclear transition energy excludes the implementation of metrology schemes that aim at a nuclear frequency standard based on ^{229}Th.

The search for the exact transition frequency in ^{229}Th resembles in a way the Hollywood biographical motion picture *Catch Me if You Can* about an FBI bank fraud agent who tracked down and finally caught a very smart criminal. A nuclear physics version of this story is taking place at present involving a number of well-established laboratories worldwide. Over three decades, physicists have attempted to "catch" the nuclear first excited state of ^{229}Th. With many indirect and inaccurate measurements, physicists so far only roughly know that the ^{229}Th first excited state may have an energy of 7.8 ± 0.5 eV above the ground state with a lifetime of around

6 hours. While such a linewidth renders ^{229}Th a valuable candidate for the next generation of time standards, i.e., nuclear clocks [58], the 1 eV uncertainty of the current measurements is too large to allow any implementation. To solve this basic problem, we propose in Chap. 5 entitled **Coherence Enhanced Optical Determination of the ^{229}Th Isomeric Transition** a method to precisely measure the transition energy using a nuclear forward scattering setup. Coherently scattered photons in the forward direction can provide a clear signature of nuclear excitation and lower the uncertainty of the transition energy down to 10 feV. Also, our proposal provides the bonus of developing nuclear quantum optics in ^{229}Th nuclei.

This thesis starts with an introduction to the theoretical treatment of light-nuclear interactions in Chap. 2, with the main focus on the so-called master equation and Maxwell-Bloch equation involving nuclear parameters. Based on the theory presented in this chapter, the three main topics discussed above are presented in the subsequent three chapters. In Chap. 3, the idea of coherent isomer triggering with fully coherent XFEL pulses is addressed. In Chap. 4, we propose two schemes to coherently control single hard x-ray photons. Finally in Chap. 5, a number of schemes to directly measure the isomeric transition energy in ^{229}Th nuclei are presented. Each of these three chapters addressing distinctive topics is organized as follows: (1) An introduction to the specific topic. Every topic has its own enchanting origin behind and we attempt to present the nice beginning of it in the introductory section, giving the flavor and also a short history for the discussed system. (2) The theoretical model and analysis of the system dynamics. This part mainly covers the theoretical model used to describe the specific physical problem. Since we address the dynamics of the system, time-scale analyses are very useful. Most of the time, we find the analysis is the "fortune teller" of a physical system as it gives a clear landscape of the whole dynamics without too involved calculations. (3) Results and discussion. In the final part of each chapter we present our numerical results and discuss the physics behind. In addition, the possible experimental implementation and available parameters are also discussed. (4) Each chapter concludes with a summary. Finally, the main conclusions and outlook are presented at the end of the thesis.

References

1. Bible Genesis 1:3
2. A.A. Penzias, R.W. Wilson, A measurement of excess antenna temperature at 4080 mc/s. Astrophys. J. **142**, 419 (1965)
3. C. Kittel, P. McEuen, *Introduction to Solid State Physics* (Wiley, New York, 1996)
4. D. Strickland, G. Mourou, Compression of amplified chirped optical pulses. Opt. Commun. **55**, 447 (1985)
5. R. Brown, R.Q. Twiss, A test of a new type of stellar interferometer on sirius. Nature **178**, 19 (1956)
6. O. Shimomura, F. Johnson, Y. Saiga, Extraction, purification and properties of aequorin, a bioluminescent protein from the luminous hydromedusan, aequorea. J. cell. comp. physiol. **59**, 223 (1962)

7. J. Livet, T. Weissman, H. Kang, J. Lu, R. Bennis, J. Sanes, J. Lichtman, Transgenic strategies for combinatorial expression of fluorescent proteins in the nervous system. Nature **450**, 56 (2007)
8. R.L. Mössbauer, Kernresonanzfluoreszenz von gammastrahlung in Ir^{191}. Zeitschrift für Physik A **151**, 124 (1958)
9. P.P. Craig, J.G. Dash, A.D. McGuire, D. Nagle, R.R. Reiswig, Nuclear resonance absorption of gamma rays in Ir^{191}. Phys. Rev. Lett. **3**, 221 (1959)
10. S.L. Ruby, Mössbauer experiments without conventional sources. J. Phys. Colloques **35**, C6–209 (1974)
11. R.L. Cohen, G.L. Miller, K.W. West, Nuclear resonance excitation by synchrotron radiation. Phys. Rev. Lett. **41**, 381 (1978)
12. G.K. Shenoy, Scientific legacy of Stanley Ruby, in *NASSAU 2006*, (Springer, Berlin Heidelberg, 2007), p. 5
13. R. Röhlsberger, *Nuclear Condensed Matter Physics with Synchrotron Radiation: Basic Principles, Methodology and Applications* (Springer-Verlag, Berlin, 2004)
14. A. Aprahamian, Y. Sun, Nuclear physics: long live isomer research. Nat. Phys. **1**, 81 (2005)
15. J. Carroll, An experimental perspective on triggered gamma emission from nuclear isomers. Laser Phys. Lett. **1**, 275 (2004)
16. D. Belic, C. Arlandini, J. Besserer, J. de Boer, J.J. Carroll, J. Enders, T. Hartmann, F. Käppeler, H. Kaiser, U. Kneissl, M. Loewe, H.J. Maier, H. Maser, P. Mohr, P. von Neumann-Cosel, A. Nord, H.H. Pitz, A. Richter, M. Schumann, S. Volz, A. Zilges, Photoactivation of $^{180}\mathrm{Ta}^{m}$ and its implications for the nucleosynthesis of nature's rarest naturally occurring isotope. Phys. Rev. Lett. **83**, 5242 (1999)
17. G.C. Baldwin, J.C. Solem, Recoilless gamma-ray lasers. Rev. Mod. Phys. **69**, 1085 (1997)
18. J.M. Blatt, V.F. Weisskopf, *Theoretical Nuclear Physics* (Courier Dover Publications, New York, 1991)
19. E. Hand, X-ray free-electron lasers fire up. Nature **461**, 708 (2009)
20. J. Amann, W. Berg, V. Blank, F. Decker, Y. Ding, P. Emma, Y. Feng, J. Frisch, D. Fritz, J. Hastings et al., Demonstration of self-seeding in a hard-x-ray free-electron laser. Nat. Photonics **6**, 693 (2012)
21. J. Feldhaus, E. Saldin, J. Schneider, E. Schneidmiller, M. Yurkov, Possible application of x-ray optical elements for reducing the spectral bandwidth of an x-ray sase fel. Opt. Commun. **140**, 341 (1997)
22. E. Saldin, E. Schneidmiller, Y. Shvyd'ko, M. Yurkov, X-ray fel with a mev bandwidth. Nucl. Instrum. Methods Phys. Res. Sect. A **475**, 357 (2001)
23. J. Arthur et al., Linac Coherent Light Source (LCLS), Conceptual Design Report (SLAC, Stanford, CA, 2002)
24. M. Altarelli et al., XFEL: The European X-Ray Free-Electron Laser, Technical Design Report (DESY, Hamburg, 2009)
25. M. Yabashi and T. Ishikawa, XFEL/SPring-8 Beamline Technical Design Report Version 2.0 (RIKEN-JASRI XFEL Project Head Office, 2010)
26. Linac coherent light source lcls-ii, SLAC, Stanford, https://slacportal.slac.stanford.edu/sites/lcls_public/lcls_ii
27. B.R. Beck, J.A. Becker, P. Beiersdorfer, G.V. Brown, K.J. Moody, J.B. Wilhelmy, F.S. Porter, C.A. Kilbourne, R.L. Kelley, Energy splitting of the ground-state doublet in the nucleus $^{229}\mathrm{Th}$. Phys. Rev. Lett. **98**, 142501 (2007)
28. Y. Shvyd'Ko, S. Stoupin, V. Blank, S. Terentyev, Near-100% bragg reflectivity of x-rays. Nat. Photonics **5**, 539 (2011)
29. F. Pfeiffer, C. David, M. Burghammer, C. Riekel, T. Salditt, Two-dimensional x-ray waveguides and point sources. Science **297**, 230 (2002)
30. A. Jarre, C. Fuhse, C. Ollinger, J. Seeger, R. Tucoulou, T. Salditt, Two-dimensional hard x-ray beam compression by combined focusing and waveguide optics. Phys. Rev. Lett. **94**, 074801 (2005)

31. S.-L. Chang, Y.P. Stetsko, M.-T. Tang, Y.-R. Lee, W.-H. Sun, M. Yabashi, T. Ishikawa, X-ray resonance in crystal cavities: realization of fabry-perot resonator for hard x rays. Phys. Rev. Lett. **94**, 174801 (2005)
32. T.J. Bürvenich, J. Evers, C.H. Keitel, Nuclear quantum optics with x-ray laser pulses. Phys. Rev. Lett. **96**, 142501 (2006)
33. I. Wong, A. Grigoriu, J. Roslund, T.-S. Ho, H. Rabitz, Laser-driven direct quantum control of nuclear excitations. Phys. Rev. A **84**, 053429 (2011)
34. W.-T. Liao, A. Pálffy, C.H. Keitel, Nuclear coherent population transfer with x-ray laser pulses. Phys. Lett. B **705**, 134 (2011)
35. Y. Shvyd'ko, S. Stoupin, A. Cunsolo, A. Said, X. Huang, High-reflectivity high-resolution x-ray crystal optics with diamonds. Nat. Phys. **6**, 196 (2010)
36. K. Liss, R. Hock, M. Gomm, B. Waibel, A. Magerl, M. Krisch, R. Tucoulou, Storage of x-ray photons in a crystal resonator. Nature **404**, 371 (2000)
37. Y.V. Shvyd'ko, M. Lerche, H.-C. Wille, E. Gerdau, M. Lucht, H.D. Rüter, E.E. Alp, R. Khachatryan, X-ray interferometry with microelectronvolt resolution. Phys. Rev. Lett. **90**, 013904 (2003)
38. T.E. Glover, M.P. Hertlein, S.H. Southworth, T.K. Allison, J. van Tilborg, E.P. Kanter, B. Krässig, H.R. Varma, B. Rude, R. Santra, A. Belkacem, L. Young, Controlling x-rays with light. Nat. Phys. **6**, 69 (2010)
39. N. Rohringer, D. Ryan, R. London, M. Purvis, F. Albert, J. Dunn, J. Bozek, C. Bostedt, A. Graf, R. Hill et al., Atomic inner-shell x-ray laser at 1.46 nanometres pumped by an x-ray free-electron laser. Nature **481**, 488 (2012)
40. P. Beiersdorfer, A.L. Osterheld, J.H. Scofield, J.R. Crespo López-Urrutia, K. Widmann, Measurement of QED and hyperfine splitting in the $2s_{1/2}$-$2p_{3/2}$ x-ray transition in Li-like $^{209}Bi^{80+}$. Phys. Rev. Lett. **80**, 3022 (1998)
41. S. Bernitt, G. Brown, J. Rudolph, R. Steinbrügge, A. Graf, M. Leutenegger, S. Epp, S. Eberle, K. Kubiček, V. Mäckel et al., An unexpectedly low oscillator strength as the origin of the fexvii emission problem. Nature **492**, 225–228 (2012)
42. Y.V. Shvyd'ko, T. Hertrich, U. van Bürck, E. Gerdau, O. Leupold, J. Metge, H.D. Rüter, S. Schwendy, G.V. Smirnov, W. Potzel, P. Schindelmann, Storage of nuclear excitation energy through magnetic switching. Phys. Rev. Lett. **77**, 3232 (1996)
43. G.V. Smirnov, U. van Bürck, J. Arthur, S.L. Popov, A.Q.R. Baron, A.I. Chumakov, S.L. Ruby, W. Potzel, G.S. Brown, Nuclear exciton echo produced by ultrasound in forward scattering of synchrotron radiation. Phys. Rev. Lett. **77**, 183 (1996)
44. W.G. Rellergert, D. DeMille, R.R. Greco, M.P. Hehlen, J.R. Torgerson, E.R. Hudson, Constraining the evolution of the fundamental constants with a solid-state optical frequency reference based on the ^{229}Th nucleus. Phys. Rev. Lett. **104**, 200802 (2010)
45. G.A. Kazakov, A.N. Litvinov, V.I. Romanenko, L.P. Yatsenko, A.V. Romanenko, M. Schreitl, G. Winkler, T. Schumm, Performance of a 229thorium solid-state nuclear clock. New J. Phys. **14**, 083019 (2012)
46. S. Matinyan, Lasers as a bridge between atomic and nuclear physics. Phys. Rep. **298**, 199 (1998)
47. A. Pálffy, C.H. Keitel, J. Evers, Single-photon entanglement in the kev regime via coherent control of nuclear forward scattering. Phys. Rev. Lett. **103**, 017401 (2009)
48. R. Coussement, Y. Rostovtsev, J. Odeurs, G. Neyens, H. Muramatsu, S. Gheysen, R. Callens, K. Vyvey, G. Kozyreff, P. Mandel, R. Shakhmuratov, O. Kocharovskaya, Controlling absorption of gamma radiation via nuclear level anticrossing. Phys. Rev. Lett. **89**, 107601 (2002)
49. R. Röhlsberger, K. Schlage, B. Sahoo, S. Couet, R. Rüffer, Collective lamb shift in single-photon superradiance. Science **328**, 1248 (2010)
50. R. Röhlsberger, H. Wille, K. Schlage, B. Sahoo, Electromagnetically induced transparency with resonant nuclei in a cavity. Nature **482**, 199 (2012)
51. O. Kocharovskaya, R. Kolesov, Y. Rostovtsev, Coherent optical control of mössbauer spectra. Phys. Rev. Lett. **82**, 3593 (1999)

52. L. Hau, S. Harris, Z. Dutton, C. Behroozi, Light speed reduction to 17 metres per second in an ultracold atomic gas. Nature **397**, 594 (1999)
53. S. Knappe, L. Liew, V. Shah, P. Schwindt, J. Moreland, L. Hollberg, J. Kitching, A microfabricated atomic clock. Appl. Phys. Lett. **85**, 1460 (2004)
54. T. Parker, Long-term comparison of caesium fountain primary frequency standards. Metrologia **47**, 1 (2009)
55. R.J. Jones, K.D. Moll, M.J. Thorpe, J. Ye, Phase-coherent frequency combs in the vacuum ultraviolet via high-harmonic generation inside a femtosecond enhancement cavity. Phys. Rev. Lett. **94**, 193201 (2005)
56. A. Cingöz, D. Yost, T. Allison, A. Ruehl, M. Fermann, I. Hartl, J. Ye, Direct frequency comb spectroscopy in the extreme ultraviolet. Nature **482**, 68 (2012)
57. S.M. Cavaletto, Z. Harman, C. Buth, C.H. Keitel, X-ray frequency combs from optically controlled resonance fluorescence. arXiv:1302.3141 (2013)
58. E. Peik and C. Tamm, Nuclear laser spectroscopy of the 3.5 ev transition in ^{229}Th. Europhys. Lett. **61**, 181 (2003)
59. C.J. Campbell, A.G. Radnaev, A. Kuzmich, V.A. Dzuba, V.V. Flambaum, A. Derevianko, Single-ion nuclear clock for metrology at the 19th decimal place. Phys. Rev. Lett. **108**, 120802 (2012)
60. Y. Shvyd'ko, *X-Ray Optics: High-Energy-Resolution Applications* (Springer-Verlag, Berlin, 2004)

Chapter 2
Theoretical Model

In this chapter, the theoretical models used in the thesis are briefly introduced. The system composed of a two-level quantum resonator, e.g., nucleus or atom interacting with a single mode of the electromagnetic wave is a basic but nontrivial problem. Many important properties of matter-field interaction can be extracted from it. For this reason we start with an example involving the coupling of a closed two-level nucleus interacting with a single-mode radiation field. We first present the used equations of motion for density matrices in Sect. 2.1. The main purpose of Sect. 2.1 is to derive the form of the Hamiltonian matrix elements adopted in nuclear physics. Furthermore, by coupling the equations for the density matrix to the Maxwell equations (i.e, Maxwell-Bloch equation) as shown in Sect. 2.2, one can study the fruitful physics of the propagation of a light pulse through a resonant medium. This issue focuses on solving the dynamics of the incident electromagnetic wave which is associated with probing a material sample from the the photons scattered off a target, e.g., as in the nuclear forward scattering setup [1]. We present the used theory in Sect. 2.2 with a very useful example about the interaction of a short light pulse with two-level nuclei. In Sect. 2.2.1, an analytic solution of the dynamics of the incident pulse is derived via Maxwell-Bloch equations confirming an expression frequently appearing in the corresponding literature as a result of a different theoretical approach. Finally, in Sect. 2.3 we demonstrate an extension of the used theory towards a three-level Λ-type system [2] which is well known for some topics in atomic quantum optics, e.g., stimulated Raman adiabatic passage (STIRAP) [3] and electromagnetically induced transparency (EIT) [4–6].

2.1 Master Equation

In this thesis, the master equation[1] is used to describe the dynamics of the considered nuclei-radiation system:

---

[1] Also called optical Bloch equation or Liouville-von Neumann equation in literature [2].

W.-T. Liao, *Coherent Control of Nuclei and X-Rays*, Springer Theses,
DOI: 10.1007/978-3-319-02120-1_2,

$$\partial_t \widehat{\rho} = \frac{1}{i\hbar} \left[\widehat{H}, \widehat{\rho}\right] + \widehat{\rho}_s. \tag{2.1}$$

Equation (2.1) describes the quantum time evolution of the density operator $\widehat{\rho}(t)$ of matter in a system, e.g., nuclei in this thesis. $\widehat{H}(t)$ is the interaction Hamiltonian between the matter and the external fields, e.g., incident electromagnetic fields, and $\widehat{\rho}_s$ describes the decoherence processes such as spontaneous decay. Moreover, Eq. (2.1) is very general for any quantum system, and the way to calculate the matrix elements of $\widehat{H}(t)$ depends on the considered physical problems. In the following, we present an example for a two-level system, and the form of $\widehat{H}(t)$ matrix elements used in nuclear physics.

A typical two-level system is illustrated in Fig. 2.1. The interaction between a nucleus and a pump laser is illustrated in Fig. 2.1c Here, pump laser (blue arrow) drives the transition $|2\rangle \leftrightarrow |1\rangle$, with detuning Δ_p. The explicit form of $\widehat{\rho}$ is:

$$\widehat{\rho} = \begin{pmatrix} \rho_{11} & \rho_{12} \\ \rho_{21} & \rho_{22} \end{pmatrix}, \tag{2.2}$$

for a two-level nuclear wavefunction $|\psi\rangle = C_1(t)|1\rangle + C_2(t)|2\rangle$

$$\rho_{eg} = C_e C_g^*. \tag{2.3}$$

Here indices $e, g \in \{1, 2\}$. Considering the spontaneous decay for the nuclear system, the decoherence matrix is

$$\widehat{\rho}_s = \frac{\Gamma}{2} \begin{pmatrix} 2\rho_{22} & -\rho_{12} \\ -\rho_{21} & -2\rho_{22} \end{pmatrix}. \tag{2.4}$$

In Eq. (2.4), the relation $\rho_{11} + \rho_{22} = 1$ is satisfied due to the conservation of population in a closed two-level system. The decay rate $\Gamma/2$ for each off diagonal coherence is derived from Eq. (2.3). Without any external laser, $\rho_{22}(t) \sim e^{-\Gamma t}$ which means $C_2(t) \sim e^{-(\Gamma/2)t}$, whence the other coherence ρ_{21} and ρ_{12} are also proportional to $e^{-(\Gamma/2)t}$. Furthermore, in the interaction picture, the interaction Hamiltonian matrix $\widehat{H}(t)$ is [3]:

$$\widehat{H} = -\frac{\hbar}{2} \begin{pmatrix} 0 & \Omega_p^* \\ \Omega_p & 2\Delta_p \end{pmatrix}, \tag{2.5}$$

where $\hbar$ is the reduced Planck constant, and Δ_p is the laser detuning. The explicit form of Eq. (2.1) can be obtained by substituting Eqs. (2.3), (2.4) and (2.5) into Eq. (2.1):

$$\partial_t \rho_{11} = \Gamma \rho_{22} + \frac{i}{2} \left(\Omega_p^* \rho_{21} - \Omega_p \rho_{21}^*\right) \tag{2.6}$$

$$\partial_t \rho_{21} = -\left(\frac{\Gamma}{2} + i\Delta_p\right) \rho_{21} - \frac{i}{2} \Omega_p \left(\rho_{22} - \rho_{11}\right) \tag{2.7}$$

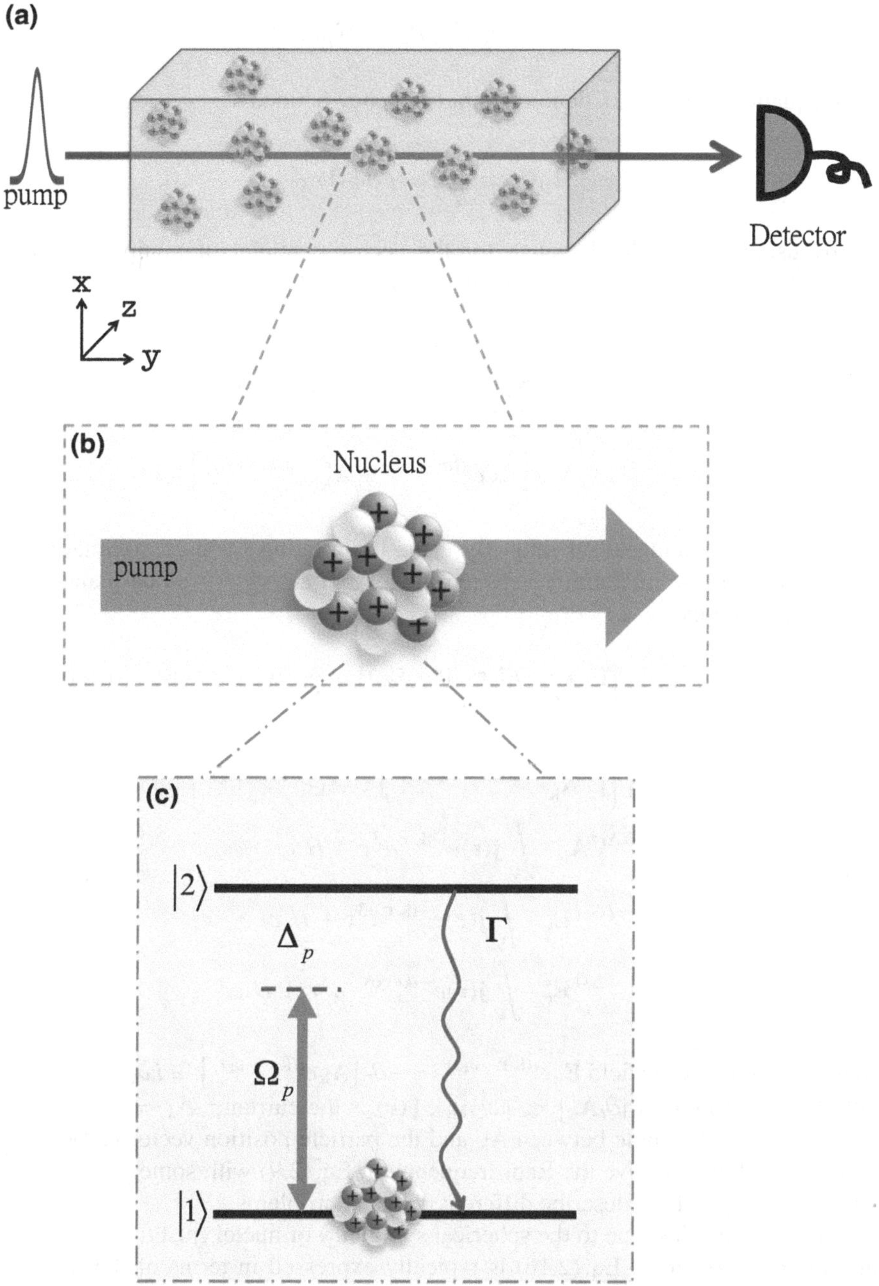

Fig. 2.1 **a** Illustration of coherent pulse propagation through a resonant medium. **b** Interaction between a nucleus and pump laser (*blue wide arrow*). **c** Sketch of a two-level nuclear system. The *blue* (Ω_p) *arrow* depicts the pump laser, and Δ_p denotes the pump laser detuning. The *green wiggled arrow* illustrates the spontaneous decay of state $|2\rangle$ with a rate of Γ

$$\partial_t \rho_{22} = -\Gamma \rho_{22} + \frac{i}{2}\left(\Omega_p \rho_{21}^* - \Omega_p^* \rho_{21}\right). \tag{2.8}$$

In the equations above, Ω_p denotes Rabi frequency defined as:

$$\Omega_p(t) = \frac{1}{\hbar}\langle 2|\widehat{H}_I(t)|1\rangle, \tag{2.9}$$

and by using the Coulomb gauge for the vector potential of pump laser $A = \sum_k A_k e^{i(kr-\omega_k t)}$ + H.c., the matrix element can be written as

$$\langle e|\widehat{H}_I(t)|g\rangle = -\langle e|\widehat{\mathbf{j}}\cdot\mathbf{A}(t)|g\rangle \tag{2.10}$$

$$= -\sum_k \langle e|\left[\widehat{\mathbf{j}}e^{-i\omega_{eg}t} + \widehat{\mathbf{j}}^* e^{i\omega_{eg}t}\right] \cdot \left[\mathbf{A_k}e^{i(\mathbf{kr}-\omega_k t)} + \mathbf{A_k^*}e^{-i(\mathbf{kr}-\omega_k t)}\right]|g\rangle \tag{2.11}$$

where $\widehat{\mathbf{j}}$ is nuclear current operator. By using the rotating wave approximation [2] and considering the interaction between nuclei and a single k mode plane wave, Eq. (2.11) becomes

$$\langle e|\widehat{H}_I(t)|g\rangle = -\langle e|\left\{\widehat{\mathbf{j}}\cdot\mathbf{A_k^*}e^{-i[\mathbf{k}\cdot\mathbf{r}+(\Omega_{eg}-\Omega_k)t]} + \widehat{\mathbf{j}}^*\cdot\mathbf{A_k}e^{i[\mathbf{k}\cdot\mathbf{r}+(\omega_{eg}-\omega_k)t]}\right\}|g\rangle \tag{2.12}$$

$$= -\langle e|\left\{\widehat{\mathbf{j}}\cdot\mathbf{A_k^*}e^{-i(\mathbf{k}\cdot\mathbf{r}-\Delta_\mathbf{k}t)} + \widehat{\mathbf{j}}^*\cdot\mathbf{A_k}e^{i(\mathbf{k}\cdot\mathbf{r}-\Delta_\mathbf{k}t)}\right\}|g\rangle \tag{2.13}$$

$$= -e^{-i\Delta_k t}\mathbf{A_k^*}\cdot\int_V \mathbf{j}(\mathbf{r})e^{-i\mathbf{k}\cdot\mathbf{r}}d^3\mathbf{r} + H.c. \tag{2.14}$$

$$= \frac{i}{\omega_k}e^{-i\Delta_k t}\mathbf{E_k^*}\cdot\int_V \mathbf{j}(\mathbf{r})e^{-i\mathbf{k}\cdot\mathbf{r}}d^3\mathbf{r} + H.c. \tag{2.15}$$

$$= \frac{i}{\omega_k}e^{-i\Delta_k t}\mathbf{E}_k^*\cdot\int_V \mathbf{j}(\mathbf{r})e^{-ikr\cos\beta}d^3r + H.c.. \tag{2.16}$$

Here, the laser electric field $\mathbf{E_k}e^{i(\mathbf{k}\cdot\mathbf{r}-\omega_k t)} = -\partial_t\left[\mathbf{A_k}e^{i(\mathbf{k}\cdot\mathbf{r}-\omega_k t)}\right] \cong i\omega_k\mathbf{A_k}e^{i(\mathbf{k}\cdot\mathbf{r}-\omega_k t)}$ with the assumption of $|\partial_t\mathbf{A_k}| \ll |\omega_k\mathbf{A_k}|$, $\mathbf{j}(\mathbf{r})$ is the current,[2] $\Delta_k = \Delta_p$ the laser detuning and β the angle between $\mathbf{A_k}$ and the particle position vector $\mathbf{r}$. Typically, the major task is to derive the Rabi frequency in Eq. (2.9) with some particular $\widehat{H}_I$ when using Eq. (2.1) to describe different physical problems.

In nuclear physics, due to the spherical symmetry of nuclei (just like atoms), the nuclear Hamiltonian of Eq. (2.16) is typically expressed in terms of the so called vector spherical harmonics $Y_{L\ell}^M(\theta, \phi)$ [7]. To derive a useful form for Eq. (2.16), we

---

[2] The current $j(r) = -i\frac{\wp\hbar}{2m}\left[\psi_\mathbf{e}^*(\mathbf{r})\nabla\psi_\mathbf{g}(\mathbf{r}) - \psi_\mathbf{g}(\mathbf{r})\nabla\psi_\mathbf{e}^*(\mathbf{r})\right]$, and the density $\rho(\mathbf{r}) = \wp\psi_\mathbf{e}^*\psi_\mathbf{g}$, where $\wp$ is the charge and m is mass of the charged particle [7].

follow the steps in Ref. [7, 8] to expand the plane wave $\mathbf{E}_k e^{ik\cos\beta}$ into multipole fields. The vector spherical harmonics $\mathbf{Y}_{L\ell}^{M}(\theta,\phi)$ is introduced as following [7]:

$$\mathbf{Y}_{L\ell}^{M}(\theta,\phi) \equiv \sum_{m=-\ell}^{\ell}\sum_{n=-1}^{1} \mathrm{C}_{\ell 1}(L,M;m,n)Y_{\ell m}(\theta,\phi)\widehat{\chi}_n \tag{2.17}$$

$$\widehat{\chi}_1 = -\frac{1}{\sqrt{2}}(\widehat{e}_x + i\widehat{e}_y) \tag{2.18}$$

$$\widehat{\chi}_0 = \widehat{e}_z \tag{2.19}$$

$$\widehat{\chi}_{-1} = \frac{1}{\sqrt{2}}(\widehat{e}_x - i\widehat{e}_y), \tag{2.20}$$

where $M = m + n$, and $Y_{\ell m}$ and $\mathrm{C}_{\ell 1}(L, M; m, n)$ are the scalar spherical harmonics and the Clebsch-Gordan coefficient (as defined in [7] or see Table A.1 in Appendix A), respectively, and $\widehat{e}_\mu$ ($\widehat{\chi}_\mu$) is the unit vector in Cartesian (spherical) coordinates. For an arbitrary vector function $R(r)\mathbf{V}(\theta,\phi)$, it can be expanded into a series:

$$\mathbf{V}(\theta,\phi) = \sum_{L=0}^{\infty}\sum_{M=-L}^{L}\sum_{\ell=L-1}^{L+1} q(L,M,\ell)\mathbf{Y}_{L\ell}^{M}(\theta,\phi) \tag{2.21}$$

$$= \sum_{L=0}^{\infty}\sum_{M=-L}^{L} \left[f(L,M)\mathbf{Y}_{LL}^{M} + g(L,M)\mathbf{Y}_{LL+1}^{M} + h(L,M)\mathbf{Y}_{LL-1}^{M}\right] \tag{2.22}$$

$$f(L,M) = \int_0^{2\pi}\int_0^{\pi} \mathbf{Y}_{LL}^{M*}(\theta,\phi)\cdot\mathbf{V}(\theta,\phi)\sin\theta d\theta d\phi \tag{2.23}$$

$$g(L,M) = \int_0^{2\pi}\int_0^{\pi} \mathbf{Y}_{LL+1}^{M*}(\theta,\phi)\cdot\mathbf{V}(\theta,\phi)\sin\theta d\theta d\phi \tag{2.24}$$

$$h(L,M) = \int_0^{2\pi}\int_0^{\pi} \mathbf{Y}_{LL-1}^{M*}(\theta,\phi)\cdot\mathbf{V}(\theta,\phi)\sin\theta d\theta d\phi, \tag{2.25}$$

First, we present the expansion of a plane wave into spherical waves:

$$e^{i\mathbf{k}\cdot\mathbf{r}} = e^{ikr\cos\beta} \tag{2.26}$$

$$= \sum_{L=0}^{\infty} Y_{L0}(\beta)\int_0^{2\pi}\int_0^{\pi} Y_{L0}^{*}(\theta)\, e^{-ikr\cos\theta}\sin\theta d\theta d\phi \tag{2.27}$$

$$= \sum_{L=0}^{\infty}(i)^L j_L(kr)\sqrt{4\pi(2L+1)}Y_{L0}(\beta) \tag{2.28}$$

$$= \sum_{L=0}^{\infty} R(L; r) Y_{L0}(\beta) \tag{2.29}$$

where $j_1(z)$ is the spherical Bessel function of first kind. Second, Eq. (2.29) shows that the field to be transformed is proportional to $e^{ikr\cos\beta}\delta_{\pm 1n}\widehat{\chi}_n$, and in the following we use Eqs. (2.17–2.25) to express it in terms of vector spherical harmonics.

$$e^{ikr\cos\beta}\widehat{\chi}_n = \sum_{L'=0}^{\infty}\sum_{M=-L'}^{L'} \big[f_n(L', M; r)\mathbf{Y}^M_{L'L'} + g_n(L', M; r)\mathbf{Y}^M_{L'L'+1} + h_n(L', M; r)\mathbf{Y}^M_{L'L'-1} \big] . \tag{2.30}$$

The coefficients f_n, g_n and h_n are derived in Appendix A, so that we merely quote the results here

$$f_n(L', M; r) = \sum_{L=0}^{\infty} R(L; r) \int_0^{2\pi}\int_0^{\pi} \mathbf{Y}^{M*}_{L'L'}(\beta, \phi) \cdot Y_{L0}(\beta)\widehat{\chi}_n \sin\beta d\beta d\phi \tag{2.31}$$

$$= \frac{1}{\sqrt{2}} R(L'; r) \left(\delta_{-1M}\widehat{\chi}^*_{-1} \cdot \widehat{\chi}_n - \delta_{1M}\widehat{\chi}^*_1 \cdot \widehat{\chi}_n \right) , \tag{2.32}$$

$$g_n(L', M; r) = \sum_{L=0}^{\infty} R(L; r) \int_0^{2\pi}\int_0^{\pi} \mathbf{Y}^{M*}_{L'L'+1}(\boldsymbol{\beta}, \boldsymbol{\phi}) \cdot Y_{L0}(\boldsymbol{\beta})\widehat{\chi}_n \sin\boldsymbol{\beta} d\boldsymbol{\beta} d\phi \tag{2.33}$$

$$= R(L'+1; r) \left[\delta_{-1M}\widehat{\chi}^*_{-1} \cdot \widehat{\chi}_n \sqrt{\frac{L'}{2(2L'+3)}} + \delta_{1M}\widehat{\chi}^*_1 \cdot \widehat{\chi}_n \sqrt{\frac{L'}{2(2L'+3)}} \right] , \tag{2.34}$$

$$h_n(L', M; r) = \sum_{L=0}^{\infty} R(L; r) \int_0^{2\pi}\int_0^{\pi} \mathbf{Y}^{M*}_{L'L'-1}(\boldsymbol{\beta}, \boldsymbol{\phi}) \cdot Y_{L0}(\boldsymbol{\beta})\widehat{\chi}_n \sin\boldsymbol{\beta} d\boldsymbol{\beta} d\phi \tag{2.35}$$

$$= R(L'-1; r) \left[\delta_{-1M}\widehat{\chi}^*_{-1} \cdot \widehat{\chi}_n \sqrt{\frac{L'+1}{2(2L'-1)}} + \delta_{1M}\widehat{\chi}^*_1 \cdot \widehat{\chi}_n \sqrt{\frac{L'+1}{2(2L'-1)}} \right] . \tag{2.36}$$

In the following, we write down the explicit form of a plane wave in terms of vector spherical harmonics:

$$\mathbf{E}_k e^{ikr\cos\beta} = \sqrt{2\pi} \sum_{n=-1,1} \left(\mathbf{E}_k \cdot \widehat{\chi}_n^*\right) \sum_{L'} (i)^{L'} \sqrt{(2L'+1)} \left(\widehat{\chi}_{-1}^* \cdot \widehat{\chi}_n \ , \ \widehat{\chi}_1^* \cdot \widehat{\chi}_n\right)$$
$$\cdot \begin{pmatrix} j_{L'}(kr)\mathbf{Y}_{L'L'}^{-1} + i\sqrt{\frac{L'}{2L'+1}} j_{L'+1}(kr)\mathbf{Y}_{L'L'+1}^{-1} - i\sqrt{\frac{L'+1}{2L'+1}} j_{L'-1}(kr)\mathbf{Y}_{L'L'-1}^{-1} \\ -j_{L'}(kr)\mathbf{Y}_{L'L'}^{1} + i\sqrt{\frac{L'}{2L'+1}} j_{L'+1}(kr)\mathbf{Y}_{L'L'+1}^{1} - i\sqrt{\frac{L'+1}{2L'+1}} j_{L'-1}(kr)\mathbf{Y}_{L'L'-1}^{1} \end{pmatrix}. \tag{2.37}$$

This can be further simplified by replacing the $\mathbf{Y}_{L'L'+1}^{\pm1}$ and $\mathbf{Y}_{L'L'-1}^{\pm1}$ with taking into account the following relation [9]

$$\frac{i}{k}\nabla\times\left[j_{L'}(kr)\mathbf{Y}_{L'L'}^{M}\right] = \sqrt{\frac{L'}{2L'+1}} j_{L'+1}(kr)\mathbf{Y}_{L'L'+1}^{M} - \sqrt{\frac{L'+1}{2L'+1}} j_{L'-1}(kr)\mathbf{Y}_{L'L'-1}^{M}. \tag{2.38}$$

Then Eq. (2.37) becomes

$$\mathbf{E}_k e^{ikr\cos\beta} = \sqrt{2\pi} \sum_{n=-1,1} \left(\mathbf{E}_k \cdot \widehat{\chi}_n^*\right) \left(\widehat{\chi}_{-1}^* \cdot \widehat{\chi}_n \ , \ \widehat{\chi}_1^* \cdot \widehat{\chi}_n\right)$$
$$\cdot \begin{pmatrix} \left(1 - \frac{1}{k}\nabla\times\right) \sum_{L'} (i)^{L'} \sqrt{(2L'+1)} j_{L'}(kr)\mathbf{Y}_{L'L'}^{-1} \\ \left(-1 - \frac{1}{k}\nabla\times\right) \sum_{L'} (i)^{L'} \sqrt{(2L'+1)} j_{L'}(kr)\mathbf{Y}_{L'L'}^{1} \end{pmatrix} \tag{2.39}$$

Finally, we substitute Eq. (2.39) into Eq. (2.16) and use the following relation [7]

$$\mathbf{Y}_{LL}^{M}(\beta,\phi) = \frac{-i(\mathbf{r}\times\nabla)Y_{LM}(\beta,\phi)}{\sqrt{L(L+1)}}, \tag{2.40}$$

together with the definition of the nuclear electric multipole moment [7]:

$$\mathbb{Q}_{LM} = \int_V r^L Y_{LM}^* \, \rho(\mathbf{r}) d^3\mathbf{r} \tag{2.41}$$

and the magnetic multipole moment [8, 9]:

$$\mathbb{M}_{LM} = \frac{1}{c(L+1)} \int_V [\mathbf{r}\times\mathbf{j}(\mathbf{r})] \cdot \nabla\left(r^L Y_{LM}^*\right) d^3\mathbf{r} \tag{2.42}$$

and the definition of the reduced transition probability $\mathbb{B}$

$$\mathbb{B}(\varepsilon L) = \frac{1}{2I_g+1} |\langle I_e \| \widehat{\mathbb{Q}}_L \| I_g \rangle|^2 \tag{2.43}$$

$$\mathbb{B}(\mu L) = \frac{1}{2I_g+1} |\langle I_e \| \widehat{\mathbb{M}}_L \| I_g \rangle|^2. \tag{2.44}$$

We obtain the general formulas for nuclear Rabi frequency and present the main steps in Appendix A, merely quoting here the results for electric/magnetic transitions [8]

$$\langle I_e, M_e|\widehat{H}_I|I_g, M_g\rangle \sim E_k\sqrt{2\pi}\sqrt{\frac{L+1}{L}}\frac{k^{L-1}}{(2L+1)!!}\mathrm{C}_{I_gI_e}(L, M; M_g, M_g) \times \sqrt{2I_g+1}\sqrt{\mathbb{B}(\varepsilon/\mu\, L)}, \quad (2.45)$$

Here the excited state $|e\rangle$ and the ground state $|g\rangle$ are characterized by the angular momenta I_e and I_g, respectively, including their magnetic sublevels M_e and M_g. Explicitly, Eq. (2.9) reads

$$\Omega_p(t) = \frac{1}{\hbar}\langle 2|\widehat{H}_I|1\rangle = \frac{4}{\hbar}\sqrt{\frac{\pi I_p(t)}{c\epsilon_0}}\sqrt{\frac{(2I_g+1)(L+1)}{L}}\frac{k_{21}^{L-1}}{(2L+1)!!}\sqrt{\mathbb{B}(\varepsilon/\mu\, L)}, \quad (2.46)$$

where $I_p(t)$ the intensity of the pump pulse, L the multipolarity of the corresponding nuclear transition and k_{21} the wave number of the corresponding nuclear transition.

Equation (2.46) is used to calculate Rabi frequencies throughout this thesis. We emphasize that the most important parameter $\mathbb{B}(\varepsilon/\mu\, L)$ characterizing the strength of the nucleus-radiation interaction is obtained from the experimental data, e.g., the Nuclear Structure and Decay Databases [10], such that no first principle calculation involving specific nuclear models is needed.

2.2 Maxwell-Bloch Equation

In some experiments, measuring the light signal scattered off a target is the main method to study the interaction between light and matter. In Chaps. 4 and 5 of this thesis, we will encounter such systems, for which the description of only Eq. (2.1) is not complete. To describe the dynamics for both matter and radiation, the coupled Maxwell-Bloch equations[3] must be used [2]:

$$\partial_t\widehat{\rho} = \frac{1}{i\hbar}\left[\widehat{H}, \widehat{\rho}\right] + \widehat{\rho}_s, \quad (2.47)$$

$$\frac{1}{c}\partial_t\Omega + \partial_y\Omega = i\eta\rho_{eg}. \quad (2.48)$$

Equation (2.48) describes the propagation of electromagnetic waves in the forward direction and is derived from Maxwell equations (see Appendix A for the derivation). The backward wave is neglected, because it is not observed in the considered problems. Furthermore, the right hand side of Eq. (2.48) is the source term associated

[3] Also called Maxwell-Schrödinger equations in literature.

with a current or a dipole moment of transition $|e\rangle \leftrightarrow |g\rangle$. The source term $i\eta\rho_{eg}$ will affect the transmission behavior of the incident radiation, where η gives the number of nuclei that may scatter the incident photons through the optical path. In Chap. 3, we will solve only Eq. (2.47) since the number of the resonant photons is much greater than the number of nuclei. For the cases treated in Chaps. 4 and 5, the situation is the other way around, the position of Eq. (2.48) becomes important.

2.2.1 Coherent Pulse Propagation Through a Resonant Medium

In this section, we will discuss the system depicted in Fig. 2.1 which is the underlying physical system discussed in for Chaps. 4 and 5. In Fig. 2.1a, a pump pulse propagates through a medium with a length L and interacts with each individual non-mutually interacting nucleus inside the medium as showed in Fig. 2.1a, b. The goal of this example is to calculate the temporal shape of the penetrating pump pulse measured by the detector placed in the forward direction. We consider a two-level nucleus described by a ground state $|1\rangle$ and an excited state $|2\rangle$, and the interaction strength between pump laser and a nucleus is given by Ω_p with a laser detuning Δ_p as depicted in Fig. 2.1c.

The theoretical model describing the considered system can be directly obtained from Eqs. (2.6), (2.7), (2.8) and (2.48).

$$\partial_t \rho_{11} = \Gamma\rho_{22} + \frac{i}{2}\left(\Omega_p^* \rho_{21} - \Omega_p \rho_{21}^*\right), \tag{2.49}$$

$$\partial_t \rho_{21} = -\left(\frac{\Gamma}{2} + i\Delta_p\right)\rho_{21} - \frac{i}{2}\Omega_p\left(\rho_{22} - \rho_{11}\right), \tag{2.50}$$

$$\partial_t \rho_{22} = -\Gamma\rho_{22} + \frac{i}{2}\left(\Omega_p \rho_{21}^* - \Omega_p^* \rho_{21}\right), \tag{2.51}$$

and

$$\frac{1}{c}\partial_t \Omega_p + \partial_y \Omega_p = i\eta\rho_{21}. \tag{2.52}$$

The initial and the boundary conditions are:

$$\rho_{eg}(0) = \delta_{1e}\delta_{g1}, \tag{2.53}$$

$$\Omega_p(0, y) = 0, \tag{2.54}$$

$$\Omega_p(t, 0) = \delta(t - \tau), \tag{2.55}$$

where indices $e, g \in \{1, 2\}$. We make here two assumptions: (1) $\Delta_p = 0$ for a resonant pump laser. (2) $\Omega_p \ll \Gamma$ for no Rabi oscillation (i.e., Ω_p is a perturbation). First, we have to linearize Eqs. (2.49), (2.50) and (2.51). By considering the assumption (2), and substituting

$$\rho_{eg}(t) \rightarrow \delta_{1e}\delta_{g1} + \kappa\rho_{eg}(t), \tag{2.56}$$

$$\Omega_p(t) \rightarrow \kappa\Omega_p(t) \tag{2.57}$$

into Eqs. (2.49)–(2.52), we obtain

$$\kappa\partial_t\rho_{11} = \Gamma\kappa\rho_{22} + \frac{i}{2}\kappa^2\left(\Omega_p^*\rho_{21} - \Omega_p\rho_{21}^*\right), \tag{2.58}$$

$$\kappa\partial_t\rho_{21} = -\frac{\Gamma}{2}\kappa\rho_{21} + \frac{i}{2}\kappa\Omega_p - \frac{i}{2}\kappa^2\Omega_p\left(\rho_{22} - \rho_{11}\right), \tag{2.59}$$

$$\kappa\partial_t\rho_{22} = -\Gamma\kappa\rho_{22} + \frac{i}{2}\kappa^2\left(\Omega_p\rho_{21}^* - \Omega_p^*\rho_{21}\right), \tag{2.60}$$

and

$$\frac{\kappa}{c}\partial_t\Omega_p + \kappa\partial_y\Omega_p = i\eta\kappa\rho_{21}. \tag{2.61}$$

Neglecting the second order κ^2 terms and subsequently using $\kappa = 1$, we obtain

$$\partial_t\rho_{11} = \Gamma\rho_{22}, \tag{2.62}$$

$$\partial_t\rho_{21} = -\frac{\Gamma}{2}\rho_{21} + \frac{i}{2}\Omega_p, \tag{2.63}$$

$$\partial_t\rho_{22} = -\Gamma\rho_{22}, \tag{2.64}$$

and

$$\frac{1}{c}\partial_t\Omega_p + \partial_y\Omega_p = i\eta\rho_{21}. \tag{2.65}$$

Thus, the dominating equations are

$$\partial_t\rho_{21} = -\frac{\Gamma}{2}\rho_{21} + \frac{i}{2}\Omega_p, \tag{2.66}$$

$$\frac{1}{c}\partial_t\Omega_p + \partial_y\Omega_p = i\eta\rho_{21}. \tag{2.67}$$

By substituting

$$\rho_{21} \rightarrow \Phi e^{-\frac{\Gamma}{2}t}, \tag{2.68}$$

$$\partial_t\rho_{21} \rightarrow -\frac{\Gamma}{2}\Phi e^{-\frac{\Gamma}{2}t} + e^{-\frac{\Gamma}{2}t}\partial_t\Phi, \tag{2.69}$$

$$\Omega_p \rightarrow Ae^{-\frac{\Gamma}{2}t}, \tag{2.70}$$

$$\partial_t\Omega_p \rightarrow -\frac{\Gamma}{2}Ae^{-\frac{\Gamma}{2}t} + e^{-\frac{\Gamma}{2}t}\partial_t A, \tag{2.71}$$

into Eqs. (2.66) and (2.67), we obtain

$$\partial_t \Phi = \frac{i}{2} A, \tag{2.72}$$

$$\frac{1}{c}\left(-\frac{\Gamma}{2} A + \partial_t A\right) + \partial_y A = i\eta\Phi. \tag{2.73}$$

By using the Fourier transform

$$\Phi(t, y) = \frac{1}{\sqrt{2\pi}} \int_{-\infty}^{\infty} \phi(\omega, k) e^{i[ky-\omega(t-\tau)]} \tag{2.74}$$

$$A(t, y) = \frac{1}{\sqrt{2\pi}} \int_{-\infty}^{\infty} \alpha(\omega, k) e^{i[ky-\omega(t-\tau)]}, \tag{2.75}$$

and substituting

$$\Phi \to \phi, \tag{2.76}$$

$$\partial_t \Phi \to -i\omega\phi, \tag{2.77}$$

$$A \to \alpha, \tag{2.78}$$

$$\partial_t A \to -i\omega\alpha, \tag{2.79}$$

$$\partial_y A \to ik\alpha, \tag{2.80}$$

into Eqs. (2.72) and (2.73), the dispersion relation of the system is obtained

$$k(\omega) = \frac{\omega}{c} - \frac{\eta}{2\omega} - i\frac{\Gamma}{2c}. \tag{2.81}$$

By using the inverse Fourier transform, the solution of α is obtained

$$A\,(t, y) = \frac{1}{\sqrt{2\pi}} \int_{-\infty}^{\infty} \alpha_0 e^{-i[k(\omega)y-\omega(t-\tau)]} d\omega \tag{2.82}$$

$$= \frac{1}{\sqrt{2\pi}} e^{-\frac{\Gamma}{2c}y} \int_{-\infty}^{\infty} \alpha_0 e^{-i[(\frac{\omega}{c}-\frac{\eta}{2\omega})y-\omega(t-\tau)]} d\omega. \tag{2.83}$$

Here, $\alpha_0 = \frac{1}{\sqrt{2\pi}} \int_{-\infty}^{\infty} \delta(t-\tau) e^{-i\omega(t-\tau)} dt = \frac{1}{\sqrt{2\pi}}$ is the Fourier transform of the boundary condition (2.55). Finally, the solution of Ω_p is (complete derivation is presented in Appendix A)

$$\Omega_p(t, y) = \frac{1}{2\pi} e^{-\frac{\Gamma}{2}[\frac{y}{c}+(t-\tau)]} \int_{-\infty}^{\infty} e^{-i[(\frac{\omega}{c}-\frac{\eta}{2\omega})y-\omega(t-\tau)]} d\omega \tag{2.84}$$

$$= \frac{1}{2\pi} e^{-\frac{\Gamma}{2}\left[\frac{y}{c}+\beta\right]} \sum_{n=0}^{\infty} \frac{(iq)^n}{n!} \int_{-\infty}^{\infty} \frac{1}{\omega^n} e^{-i\omega z} d\omega \tag{2.85}$$

$$= \frac{1}{\sqrt{2\pi}} e^{-\frac{\Gamma}{2}\left[\frac{y}{c}+\beta\right]} \sum_{n=0}^{\infty} \frac{(iq)^n}{n!} \left[-i\sqrt{\frac{\pi}{2}} \frac{(-iz)^{n-1}}{(n-1)!} sgn(z)\right] \tag{2.86}$$

$$= \delta(z) e^{-\frac{\Gamma}{2}\left[\frac{y}{c}+\beta\right]} - q \frac{J_1\left(2\sqrt{qz'}\right)}{2\sqrt{qz'}} e^{-\frac{\Gamma}{2}\left[\frac{y}{c}+\beta\right]}. \tag{2.87}$$

Here, $J_1(z)$ is the Bessel function of first kind [11, 12], $\beta = t - \tau$, $z = \frac{y}{c} - (t - \tau)$, $z' = -z$ and $sgn(z)$ is the sign function [13] which equals -1 in our case since $z < 0$. Additionally, in most cases $\frac{L}{c}$ is much smaller than $t - \tau$, i.e., the prorogation time of the incident light pulse through a target with length L is much shorter than the detection time window $t - \tau$. Therefore, the terms $\frac{y}{c}$ are typically negligible. Moreover, we derive Eq. (2.86) from Eq. (2.85) by using the Fourier transform of $1/\omega^n$. The explicit form of Eq. (2.87) is then

$$\begin{aligned} \Omega_p(t, y) &= \left\{ \delta\left[\frac{y}{c} - (t-\tau)\right] - \left(\frac{\eta y}{2}\right) \frac{J_1\left[2\sqrt{\left(\frac{\eta y}{2}\right)\left(t - \tau - \frac{y}{c}\right)}\right]}{2\sqrt{\left(\frac{\eta y}{2}\right)\left(t - \tau - \frac{y}{c}\right)}} \right\} e^{-\frac{\Gamma}{2}\left(\frac{y}{c}+t-\tau\right)} \\ &= \left\{ \delta\left[\frac{y}{c} - (t-\tau)\right] - \left(\frac{\xi\Gamma y}{4L}\right) \frac{J_1\left[2\sqrt{\left(\frac{\xi\Gamma y}{4L}\right)\left(t - \tau - \frac{y}{c}\right)}\right]}{2\sqrt{\left(\frac{\xi\Gamma y}{4L}\right)\left(t - \tau - \frac{y}{c}\right)}} \right\} e^{-\frac{\Gamma}{2}\left(\frac{y}{c}+t-\tau\right)}. \end{aligned} \tag{2.88}$$

In experiments, the measured field [1, 11, 12, 14] is proportional to

$$\Omega_p(t, L) = \left\{ \delta\left[\frac{L}{c} - (t-\tau)\right] - \left(\frac{\xi\Gamma}{4}\right) \frac{J_1\left[2\sqrt{\left(\frac{\xi\Gamma}{4}\right)\left(t - \tau - \frac{L}{c}\right)}\right]}{2\sqrt{\left(\frac{\xi\Gamma}{4}\right)\left(t - \tau - \frac{L}{c}\right)}} \right\} e^{-\frac{\Gamma}{2}\left(\frac{L}{c}+t-\tau\right)}, \tag{2.89}$$

and the measured signal is proportional to

$$|\Omega_p(t, L)|^2 = \left\{ \delta\left[\frac{L}{c} - (t-\tau)\right] - \left(\frac{\xi\Gamma}{4}\right) \frac{J_1\left[2\sqrt{\left(\frac{\xi\Gamma}{4}\right)\left(t - \tau - \frac{L}{c}\right)}\right]}{2\sqrt{\left(\frac{\xi\Gamma}{4}\right)\left(t - \tau - \frac{L}{c}\right)}} \right\}^2 e^{-\Gamma\left(\frac{L}{c}+t-\tau\right)}. \tag{2.90}$$

This equation can be derived also using an iterative method with the response function as it was shown in Ref. [15]. We will use Eq. (2.90) in Chaps. 4 and 5 to explain a phenomenon called dynamical beat [16] of nuclear forward scattering [1].

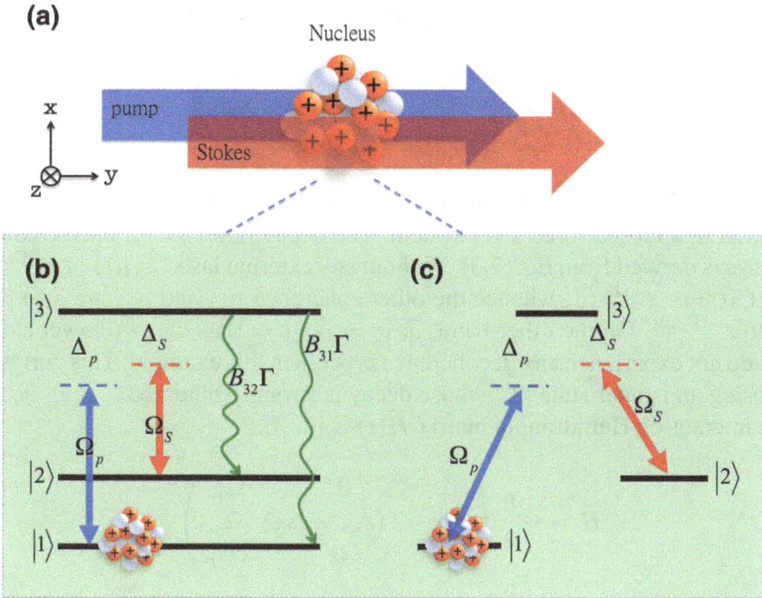

Fig. 2.2 **a** Interaction between a nucleus, pump laser (*blue wide arrow*) and Stokes laser (*red wide arrow*). Typical sketches of a three-level Λ-type system appear in the literature of **b** nuclear and **c** atomic physics. The *blue* (Ω_p) and *red* (Ω_S) *vertical arrows* depict pump and Stokes lasers, respectively. Δ_p and Δ_S denote the detunings of the corresponding lasers. The *green wiggled arrows* illustrate the spontaneous decay of state $|3\rangle$, B_{31} (B_{32}) is the branching ratio of $|3\rangle \leftrightarrow |1\rangle$ ($|3\rangle \leftrightarrow |2\rangle$) transition and Γ is the spontaneous decay rate of state $|3\rangle$

2.3 Three-Level Λ-Type System

In this section, we extend the used theory for a three-level Λ-type system [2] that is well known for describing several effects in atomic quantum optics, for example, stimulated Raman adiabatic passage (STIRAP) [3] and electromagnetically induced transparency (EIT) [4–6]. A typical three-level Λ-type system is illustrated in Fig. 2.2a. For convenience, we assume the wave function of the considered nucleus is $|\psi\rangle = C_1(t)|1\rangle + C_2(t)|2\rangle + C_3(t)|3\rangle$. The interaction between the nucleus and two lasers is typically illustrated by a sketch like Fig. 2.2b in nuclear physics or Fig. 2.2c in atomic physics. Since the pattern of Fig. 2.2c looks like the Greek letter Λ, it is called as a Λ-type system. Here, pump laser (blue arrow) drives the transition $|3\rangle \leftrightarrow |1\rangle$, with detuning Δ_p and Stokes laser (red arrow) drives the transition $|3\rangle \leftrightarrow |2\rangle$ with detuning Δ_S. The explicit form of $\widehat{\rho}$ in this case is:

$$\widehat{\rho} = \begin{pmatrix} \rho_{11} & \rho_{12} & \rho_{13} \\ \rho_{21} & \rho_{22} & \rho_{23} \\ \rho_{31} & \rho_{32} & \rho_{33} \end{pmatrix}. \tag{2.91}$$

By considering the spontaneous decay, the decoherence matrix is

$$\widehat{\rho_s} = \frac{\Gamma}{2}\begin{pmatrix} 2B_{31}\rho_{33} & 0 & -\rho_{13} \\ 0 & 2B_{32}\rho_{33} & -\rho_{23} \\ -\rho_{31} & -\rho_{32} & -2\rho_{33} \end{pmatrix}. \tag{2.92}$$

In Eq. (2.92), $B_{31} + B_{32} = 1$ due to $\rho_{11} + \rho_{22} + \rho_{33} = 1$, i.e., the conservation of population in a closed three level system. The decay rate $\Gamma/2$ for each of diagonal coherence is derived from Eq. (2.3). Without any external laser, $\rho_{33}(t) \sim e^{-\Gamma t}$ which means $C_3(t) \sim e^{-(\Gamma/2)t}$, whence the other coherence $\rho_{3\mu}$ and $\rho_{\mu 3}$ are also proportional to $e^{-(\Gamma/2)t}$. On the other hand, $\widehat{\rho_{s,12}} = \widehat{\rho_{s,21}} = 0$ as the two lower states $|1\rangle$ and $|2\rangle$ do not experience any decoherence process in this example. This corresponds to choosing an isomer state $|2\rangle$, whose decay is strongly hindered.

The interaction Hamiltonian matrix $\widehat{H}(t)$ is [2, 3]:

$$\widehat{H} = -\frac{\hbar}{2}\begin{pmatrix} 0 & 0 & \Omega_p^* \\ 0 & -2\left(\Delta_p - \Delta_S\right) & \Omega_S^* \\ \Omega_p & \Omega_S & 2\Delta_p \end{pmatrix}, \tag{2.93}$$

where $\hbar$ is the reduced Planck constant. The explicit form of Eq. (2.1) can be obtained by substituting Eqs. (2.91), (2.92) and (2.93) into Eq. (2.1):

$$\partial_t \rho_{11} = B_{31}\Gamma\rho_{33} + \frac{i}{2}\left(\Omega_p^*\rho_{31} - \Omega_p\rho_{31}^*\right) \tag{2.94}$$

$$\partial_t \rho_{21} = i\left(\Delta_S - \Delta_p\right)\rho_{21} + \frac{i}{2}\left(\Omega_S^*\rho_{31} - \Omega_p\rho_{32}^*\right) \tag{2.95}$$

$$\partial_t \rho_{31} = -\left(\frac{\Gamma}{2} + i\Delta_p\right)\rho_{31} + \frac{i}{2}\Omega_S\rho_{21} - \frac{i}{2}\Omega_p\left(\rho_{33} - \rho_{11}\right) \tag{2.96}$$

$$\partial_t \rho_{22} = B_{32}\Gamma\rho_{33} + \frac{i}{2}\left(\Omega_S^*\rho_{32} - \Omega_S\rho_{32}^*\right) \tag{2.97}$$

$$\partial_t \rho_{32} = -\left(\frac{\Gamma}{2} + i\Delta_S\right)\rho_{32} + \frac{i}{2}\Omega_p\rho_{21}^* - \frac{i}{2}\Omega_S\left(\rho_{33} - \rho_{22}\right) \tag{2.98}$$

$$\partial_t \rho_{33} = -\Gamma\rho_{33} + \frac{i}{2}\left(\Omega_p\rho_{31}^* - \Omega_p^*\rho_{31}\right) + \frac{i}{2}\left(\Omega_S\rho_{32}^* - \Omega_S^*\rho_{32}\right). \tag{2.99}$$

In above equations, Ω_p and Ω_S are Rabi frequencies defined as:

$$\begin{aligned} \Omega_p(t) &= \frac{1}{\hbar}\langle 3|\widehat{H}_I|1\rangle \\ &= \frac{4}{\hbar}\sqrt{\frac{\pi I_p(t)}{c\epsilon_0}}\sqrt{\frac{(2I_1+1)(L_{31}+1)}{L_{13}}}\frac{k_{31}^{L_{31}-1}}{(2L_{13}+1)!!}\sqrt{\mathbb{B}(\varepsilon/\mu\, L_{13})}, \end{aligned} \tag{2.100}$$

$$\Omega_S(t) = \frac{1}{\hbar}\langle 3|\widehat{H}_I|2\rangle$$

$$= \frac{4}{\hbar}\sqrt{\frac{\pi I_S(t)}{c\epsilon_0}}\sqrt{\frac{(2I_2+1)(L_{23}+1)}{L_{23}}}\frac{k_{32}^{L_{23}-1}}{(2L_{23}+1)!!}\sqrt{\mathbb{B}(\varepsilon/\mu\, L_{23})}, \quad (2.101)$$

where $I_{p(S)}(t)$ is the pump (Stokes) pulse, $I_{1(2)}$ is the angular momentum of ground state $|1(2)\rangle$, and , $L_{1(2)3}$ is the multipolarity of the corresponding nuclear transition $|1(2)\rangle \leftrightarrow |3\rangle$. Equations (2.94)–(2.101) are successfully used to explain plenty of phenomena in atomic quantum optics. In this thesis, we adopt and use it to investigate the proposal called nuclear coherent population transfer (NCPT) in Chap. 3. Furthermore, Eqs. (2.94)–(2.101) together with

$$\frac{1}{c}\partial_t\Omega_p + \partial_y\Omega_p = i\eta\rho_{31}, \quad (2.102)$$

will be used to discuss another proposal labeled as electromagnetically modified nuclear forward scattering in Chap. 5.

References

1. R. Röhlsberger, *Nuclear Condensed Matter Physics with Synchrotron Radiation: Basic Principles, Methodology and Applications* (Springer-Verlag, Berlin, 2004)
2. M.O. Scully, M.S. Zubairy, *Quantum Optics* (Cambridge University Press, Cambridge, 1997)
3. K. Bergmann, H. Theuer, B.W. Shore, Coherent population transfer among quantum states of atoms and molecules. Rev. Mod. Phys. **70**, 1003 (1998)
4. L. Hau, S. Harris, Z. Dutton, C. Behroozi, Light speed reduction to 17 metres per second in an ultracold atomic gas. Nature **397**, 594 (1999)
5. W.-H. Lin, W.-T. Liao, C.-Y. Wang, Y.-F. Lee, I.A. Yu, Low-light-level all-optical switching based on stored light pulses. Phys. Rev. A **78**, 033807 (2008)
6. Y.-W. Lin, W.-T. Liao, T. Peters, H.-C. Chou, J.-S. Wang, H.-W. Cho, P.-C. Kuan, I.A. Yu, Stationary light pulses in cold atomic media and without bragg gratings. Phys. Rev. Lett. **102**, 213601 (2009)
7. J. M. Blatt, V. F. Weisskopf, *Theoretical Nuclear Physics* (Courier Dover Publications, New York, 1991)
8. A. Pálffy, J. Evers, C.H. Keitel, Electric-dipole-forbidden nuclear transitions driven by super-intense laser fields. Phys. Rev. C **77**, 044602 (2008)
9. P. Ring, P. Schuck, *The Nuclear Many-Body Problem* (Springer Verlag, New York, 1980)
10. Nuclear structure and decay databases, http://www.nndc.bnl.gov
11. M.D. Crisp, Propagation of small-area pulses of coherent light through a resonant medium. Phys. Rev. A **1**, 1604 (1970)
12. Y.V. Shvyd'ko, U. van Bürck, W. Potzel, P. Schindelmann, E. Gerdau, O. Leupold, J. Metge, H.D. Rüter, G.V. Smirnov, Hybrid beat in nuclear forward scattering of synchrotron radiation. Phys. Rev. B **57**, 3552 (1998)
13. L. Rade, B. Westergren, *Mathematics Handbook for Science and Engineering* 4th ed. (Springer, Berlin, 2004)
14. U. V. Bürck, Coherent pulse propagation through resonant media, Hyperfine Interact. **123/124**, 483 (1999)
15. Y. Shvyd'ko, Nuclear resonant forward scattering of x rays: time and space picture. Phys. Rev. B **59**, 9132 (1999)
16. Y.V. Shvyd'ko, U. van Bürck, Hybrid forms of beat phenomena in nuclear forward scattering of synchrotron radiation. Hyperfine Interact. **123–124**, 511 (1999)

Chapter 3
Nuclear Coherent Population Transfer with X-Ray Laser Pulses

This chapter aims to address the problem of *actively manipulating the nuclear state by using coherent hard x-ray photons*. In particular, we are interested in efficient control of the nuclear population dynamics in a three level system associating to level schemes of isomer triggering, i.e., the Λ-type system.[1] Typically, the traditional way of shifting nuclei from one internal quantum state to another is done by incoherent photon absorption, i.e., incoherent γ-rays illuminate the nuclear sample and excite the nuclei to some high-energy states. Subsequently, some of the excited nuclei may decay to the target state by chance, according to the corresponding branching ratio. This kind of method is rather passive, and its efficiency is low. Encouraged by the development of X-ray Free Electron Laser (XFEL) [1–6], an improved version was considered [7] using coherent x-ray absorption. However, even this scheme stays in the traditional and passive frame and only increases perhaps the excitation efficiency. As the first nuclear quantum optics application presented in this thesis, we present another XFEL-based leap forward, coherently transferring nuclei between two states in a determined and controlled way. For the first time, the nuclear coherent population transfer (NCPT) between the two lower states in a nuclear three-level scheme is studied by utilizing two overlapping x-ray laser pulses in the stimulated Raman adiabatic passage (STIRAP) setup [8]. Since most of the nuclear transition energies are higher than the energies of the currently producible x-ray photons, we introduce the concept of an accelerated nuclear target, i.e., a nuclear beam produced by particle accelerators [9].

This chapter starts with a brief introduction on isomer triggering followed by the two methods considered to achieve this: the π-pulse method and the stimulated Raman adiabatic passage. In Sect. 3.1, we introduce the basic features of the π-pulse and of the so-called dark state which plays the central role of STIRAP. Also, four intuitive difficulties on driving nuclear transitions by x-ray lasers are discussed here, showing how vital the optimization of the laser parameters is for controlling the nuclear state. Following this conclusion, in Sect. 3.2 we present in detail the two

[1] The origin of the name "Λ-type" has been discussed in Sect. 2.3 and is due to the pattern of the sketch of the three level scheme.

W.-T. Liao, *Coherent Control of Nuclei and X-Rays*, Springer Theses,

DOI: 10.1007/978-3-319-02120-1_3,

schemes to achieve NCPT with XFEL and find out the required laser intensities. In Sect. 3.3, the possible facilities of XFEL and ion accelerators used to implement NCPT are discussed. Two more possible experimental issues regarding the influence of the apparatus errors are also addressed. In addition to the large scale infrastructures, we propose to utilize tabletop solutions, which are well developed in the field of laser-plasma interaction. This tabletop approach also deserves attention for the problem of a global experimental availability of the nuclear state manipulation.

3.1 Motivation and Introduction

Nuclear metastable states, also known as isomers, can store large amounts of energy over longer periods of time. Isomer depletion, i.e., release on demand of the energy stored in isomers, has received a lot of attention in the last one and a half decades, especially related to the fascinating prospects of nuclear batteries [10–12]. Two notable examples on triggered γ-emission from nuclear isomers by x-ray absorption are ^{180m}Ta [13–15] and ^{178m2}Hf [15–20]. As shown in Fig. 3.1, by shining only pump radiation pulse, depletion occurs when the nuclear population in isomer state $|1\rangle$ is excited to a higher triggering level $|3\rangle$ whose decay to other lower levels, e.g., state $|2\rangle$ is no longer hindered by the long-lived isomer. However, such control of nuclear state is done by incoherent processes and its efficiency is therefore low. To improve the efficiency for controlling nuclei, we apply two schemes of coherent population transfer to nuclear system with coherent x-ray beams. Coherent population transfer in the atomic nuclei would not only be a powerful tool for preparation and detection in nuclear physics, but also especially useful for control of energy stored in nuclear states. Moreover, the coherent photoactivation of nuclear isomers facilitates the understanding of the synthesis of the naturally occurring isotopes in stars [14]. We briefly introduce two schemes of coherent population transfer in the following.

In atomic and molecular physics, a successful and robust way for coherent population transfer is the STIRAP [8], a technique in which two coherent fields couple

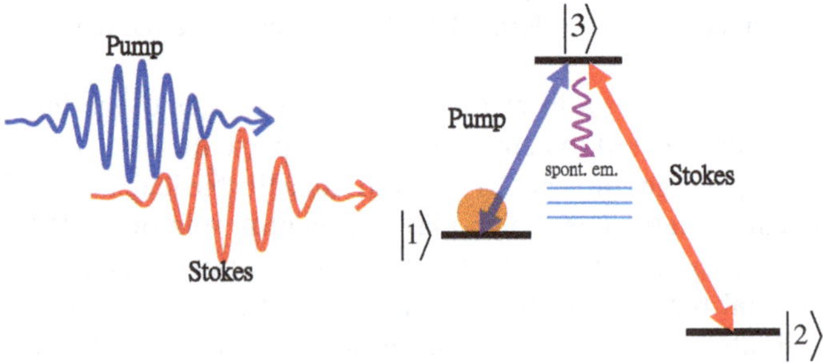

Fig. 3.1 The Λ-level scheme. The *blue arrow* illustrates the pump pulse, and the *red arrow* depicts the Stokes pulse. All initial population is in state $|1\rangle$

to a three-level system as showed in Fig. 3.1. The interaction of a Λ-level scheme with the pump laser P driving the $|1\rangle \rightarrow |3\rangle$ transition and the Stokes laser S driving the $|2\rangle \rightarrow |3\rangle$ transition is depicted in Fig. 3.1. In STIRAP, at first the Stokes laser creates a superposition of the two unpopulated states $|2\rangle$ and $|3\rangle$. Subsequently, the pump laser couples the fully occupied $|1\rangle$ and the pre-built coherence of the two empty states. In the whole process, the dark (trapped) state

$$|D\rangle = \frac{\Omega_S(t)}{\sqrt{\Omega_P^2(t) + \Omega_S^2(t)}}|1\rangle - \frac{\Omega_P(t)}{\sqrt{\Omega_P^2(t) + \Omega_S^2(t)}}|2\rangle. \quad (3.1)$$

is formed and evolves with the time dependent pump and Stokes Rabi frequencies[2] $\Omega_P(t)$ and $\Omega_S(t)$, respectively [8]. Obviously, one can control the populations in the states $|1\rangle$ and $|2\rangle$ via temporally adjusting the laser parameters, e.g., the laser electric field strengths of pump and Stokes.

Another option for achieving the coherent population transfer is utilizing so-called π-pulses. Using the scheme in Fig. 3.1, let us consider the interaction of a two-level system with a single-mode laser, driving the $|1\rangle \rightarrow |3\rangle$ transition by the pump laser (for now we temporarily neglect state $|2\rangle$ and the Stokes laser). The eigenstate of this system is [21]

$$|\psi\rangle = \cos\left(\frac{1}{2}\int_{-\infty}^{t} \Omega_P(\tau)\, d\tau\right)|1\rangle + \sin\left(\frac{1}{2}\int_{-\infty}^{t} \Omega_P(\tau)\, d\tau\right)|3\rangle. \quad (3.2)$$

Obviously, the complete coherent population transfer happens when

$$\int_{-\infty}^{t} \Omega_P(\tau)\, d\tau = n\pi \quad (3.3)$$

for n odd. Because of this particular case, $\Omega_P(\tau)$ is called a π-pulse if $\int_{-\infty}^{\infty} \Omega_P(\tau)\, d\tau = \pi$. In the scheme in Fig. 3.1, one can shine a pump π-pulse and subsequently a Stokes π-pulse on the target to coherently channel all population from state $|1\rangle$ to state $|2\rangle$ via the intermediate state $|3\rangle$. This technique is termed as two π-pulses method. One question may arise: why not directly pump the population from $|1\rangle$ to state $|2\rangle$ by using just one laser pulse? The advantages of the considered two-field scheme is emphasized by K. Bergmann et al. [8]: "*...the use of two lasers coupling three states, rather than a single laser coupling two states, offers many advantages: the excitation efficiency can be made relatively insensitive to many of the experimental details of the pulses. In addition, with the three-state system, one can produce excitation between states of the same parity, for which single-photon transitions are forbidden for electricdipole radiation, or between magnetic sublevels...*". Similarly in the case of NCPT illustrated in Fig. 3.1, either such direct transition between two ground

[2] The Rabi frequency is defined as $\Omega = \frac{1}{\hbar}\langle\psi_f|\widehat{P} \cdot E|\psi_i\rangle$. Here $\widehat{P}$ is the dipole operator, E denotes the electric field of the laser and $\hbar$ is the reduced Planck constant.

states is forbidden (e.g., the isomer state), or the required laser intensity of using one field will be higher than that of using two lasers due to the linewidth of the nuclear ground state $|2\rangle$ which may be much narrower than that of state $|3\rangle$. Because of these two reasons, a two-field NCPT is considered in this chapter.

The transfer of such schemes to nuclear systems, although encouraged by the progress of laser technology, has not been accomplished due to the lack of γ-ray laser sources. The pursuit of coherent sources for wavelengths around or below 1 nm is supported however by the advent and commissioning of XFEL [1–6], the availability of which stimulates the transfer of quantum optical schemes to nuclei. Furthermore, a nuclear three-level scheme that can lead to the depletion of a metastable state. However, four intuitive problems are on the way to extend quantum optics into the nuclear region:

1. The tiny size of the atomic nucleus. The typical size of the atomic nucleus is on the order of 10^{-14} m, or 10 fm, which results in a very small nuclear dipole moment $\mathbf{P}_N \sim 10^{-33}$ C·m, whereas the atomic dipole moment is on the order of 10^{-29} C·m. Considering the general dipole-electric field coupling $\mathbf{P}_N \cdot \mathbf{E}$, the nucleus-radiation coupling is four orders of magnitude smaller than the atom-light interaction for the same applied laser intensity.
2. The nuclear transition energy is typically greater than the photon energy from available coherent light sources. To bridge the gap between x-ray laser frequency and nuclear transition energies, a key proposal is to combine moderately accelerated target nuclei and novel x-ray lasers [9]. A nuclear beam suitably accelerated can interact with Lorentz boosted x-ray laser pulses, i.e., in the nuclear rest frame the photon frequency becomes resonant to the corresponding nuclear transition if the relativistic factor γ of the accelerated nuclei is properly chosen. This XFEL-accelerated nuclei head-on collider system will be considered throughout this chapter.
3. The nuclear linewidth is typically narrower than the laser bandwidth. This property limits the resonant photon number within a produced XFEL pulse and gives a lower effective intensity [22]. An intuitive picture of this issue is presented in Fig. 3.2. The intensity I of an incident XFEL pulse will not be fully observed by the nuclei, and the effective intensity I^{eff} depends on the ratio of the laser bandwidth to the nuclear linewidth. For instance, the effective intensity of a short incident pulse illustrated in Fig. 3.2c, d is much weaker than that of a long pulse case (Fig. 3.2a, b), because Γ is wider than the linewidth of the latter.[3] An expression of the effective pulse shape is given in Appendix B.
4. The coherence of XFEL. Both the success of STIRAP and that of two π-pulse are based on the full coherence of the laser pulses, therefore the fully coherent XFEL source is also paramount for NCPT. Thus, in the whole chapter we assume

[3] A metaphor of the effective intensity can be grasped from the experience of going to a dentist. When a patient is undertaking dental x-ray diagnosis, he/she will not see the x-ray from the machine. This is due to the fact that the frequency of the used x-ray burst is too high to be observed by the human eye. Thus, the effective intensity of the x-ray is zero for it. Just as the human eye is only reacting to visible light, and not to x-rays, nuclei only react to photons with their resonant frequencies.

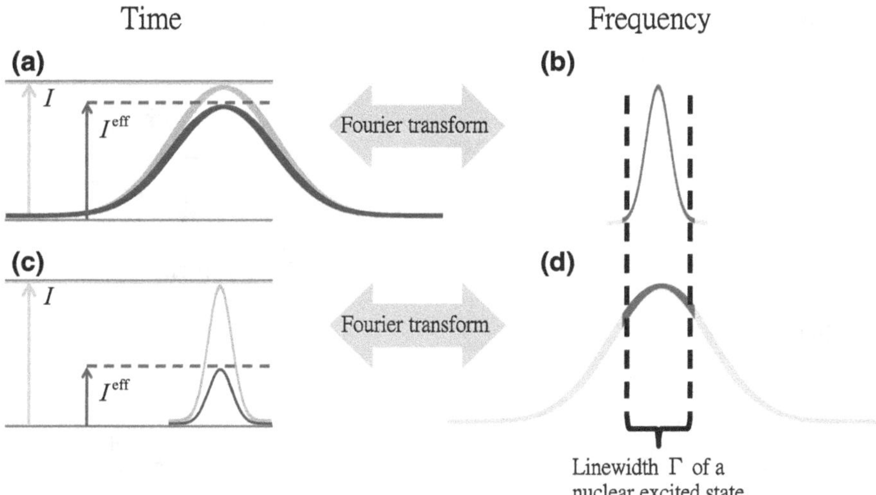

Fig. 3.2 The intuitive sketch of the concept of the effective intensity I^{eff}. For the long laser pulse case in (**a**, **b**), the bandwidth of the incident laser of intensity I is narrower than the linewidth Γ of some nuclear excited state. For the short pulse case in (**c**, **d**), the effective intensity is significantly reduced since the linewidth of the incident laser is wider than Γ

a fully coherent XFEL source such as the future XFEL Oscillator (XFELO) [23] or the seeded XFEL (SXFEL) [1–6] for both pump and Stokes lasers.

Consequently, increasing the laser intensity and optimizing the laser parameters in the nuclear rest frame become the only means of manipulating nuclear states, since the nuclear size is an intrinsic property. In the following, the fully coherent XFEL pulses are used together with acceleration of the target nuclei to achieve the resonance condition. The needed optimal laser intensity will be discussed for achieving not only 100 % NCPT but also coherent isomer triggering.

3.2 Nuclear Coherent Population Transfer

3.2.1 Model, Time Scale and Parameters

We study a collider system depicted in Fig. 3.3, composed of an accelerated nuclear beam that interacts with two incoming XFEL pulses. The explanation of the notations used throughout the following text and the equations can be found in Table 3.1. As mentioned before, the nuclear excitation energies are typically higher than the designed photon energy of the XFELO and SXFEL. The accelerated nuclei can interact with two Doppler-shifted x-ray laser pulses as showed in Fig. 3.3a. The two laser frequencies (two-color) and the relativistic factor γ of the accelerated nuclei

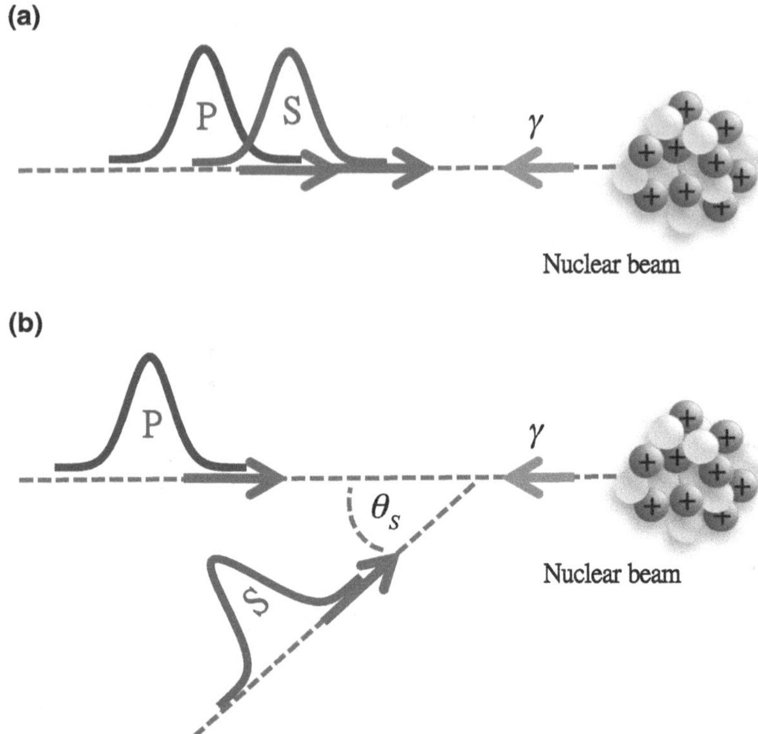

Fig. 3.3 **a** Two-*color* scheme (copropagating-beams) in the lab frame. In this setup, the frequency of pump laser is different from that of the Stokes laser. The nuclear beam is accelerated such that $\gamma(1+\beta)\omega_{p(S)} = ck_{31(2)}$ is fulfilled. **b** One-*color* scheme (crossed-beams) in the lab frame. In this setup $\omega_p = \omega_S$, and the nuclear beam is accelerated such that both conditions $\gamma(1+\beta)\omega_p = ck_{31}$ and $\gamma(1+\beta\cos\theta_S)\omega_S = ck_{32}$ are fulfilled

have to be chosen such that in the nuclear rest frame both one-photon resonances are fulfilled. Copropagating laser pulses should have different frequencies in the laboratory frame in order to match the nuclear transition energies. To fulfill the resonance conditions with a one-color laser we envisage the pump and Stokes pulses meeting the nuclear beam at different angles ($\theta_S \neq 0$), as shown in Fig. 3.3b.

In the following we address the laser beam parameter requirements. The most important prerequisite for nuclear STIRAP is the *temporal* coherence of the x-ray lasers. The coherence time of the existent XFEL at the Linac Coherent Light Source (LCLS) in Stanford, USA and of the European XFEL are on the order of 1 fs, much shorter than the pulse duration of 100 fs [3, 4, 24]. The SXFEL, considered as an upgrade for both facilities, will deliver completely transversely and temporally coherent pulses, that can reach 0.1 ps pulse duration and about 10 meV bandwidth [2, 6]. Recently, a self-seeding scheme successfully produced a near Fourier-transform-limited x-ray pulses with 0.4–0.5 eV bandwidth at 8–9 keV photon

Table 3.1 The notations used throughout the text. The indices $i, j = 1, 2, 3$ denote the three nuclear states showed in Fig. 3.4. The label 'Lab' ('Rest') indicates that the corresponding values are in the lab (nuclear rest) frame

Notation	Frame	Explanation
c	Any	Speed of light in vacuum
β	Lab	Velocity of the nuclear particle, in units of c
γ	Lab	Relativistic factor $\gamma = 1/\sqrt{1-\beta^2\cos^2\theta}$
ϵ_0	Any	Vacuum permittivity
h	Any	Planck constant, $h = 2\pi\hbar$
Γ	Rest	Spontaneous decay rate of $\|3\rangle$
ρ_{ij}	Rest	Density matrix element
δ_{ij}	Any	Kronecker delta
k_{3i}	Rest	Wave number of $\|3\rangle \rightarrow \|i\rangle$ transition
B_{3i}	Rest	Branching ratio of $\|3\rangle \rightarrow \|i\rangle$ spontaneous decay
$\Gamma_{Lp(S)}$	Lab	Laser bandwidth of pump (Stokes)
$\Omega_{p(S)}$	Rest	Slowly varying effective Rabi frequency of pump (Stokes) laser
$\tau_{p(S)}$	Rest	Temporal peak position of pump (Stokes) laser
$\omega_{p(S)}$	Lab	Angular frequency of pump (Stokes) laser
$\Delta_{p(S)}$	Rest	Laser detuning, $\Delta_{p(S)} = \gamma(1+\beta\cos\theta)\omega_{p(S)} - ck_{31(2)}$
$E_{p(S)}$	Lab	Slowly varying envelope of electric field of pump (Stokes) laser
$I_{p(S)}$	Lab	Peak intensity of pump (Stokes) laser pulse, $I_{p(S)} = \frac{1}{2}c\epsilon_0 E^2_{p(S)}$
$I^{\text{eff}}_{p(S)}$	Lab	Effective peak intensity of pump (Stokes) laser pulse $I^{\text{eff}}_{p(S)} = I_{p(S)}\frac{\Gamma}{\gamma(1+\beta\cos\theta)\Gamma_{p(S)}}$ for $\Gamma < \gamma(1+\beta\cos\theta)\Gamma_{p(S)}$ $I^{\text{eff}}_{p(S)} = I_{p(S)}$ for $\Gamma \geq \gamma(1+\beta\cos\theta)\Gamma_{p(S)}$
$T_{p(S)}$	Lab	Pulse duration of pump (Stokes) laser
$I_{1(2)}$	Any	Angular momentum of ground state $\|1\rangle$ ($\|2\rangle$)
L_{i3}	Any	Multipolarity of the corresponding nuclear $\|i\rangle \rightarrow \|3\rangle$ transition
$\mathbb{B}(\varepsilon/\mu\, L_{i3})$	Rest	Reduced transition probability for the nuclear electric (ε) or magnetic (μ) $\|i\rangle \rightarrow \|3\rangle$ transition

energy [25]. Another option is the XFELO that will provide coherence time on the order of the pulse duration ~1 ps, and meV narrow bandwidth [23]. We consider here the laser photon energy for the pump laser fixed at 25 keV for the XFELO and 12.4 keV for the SXFEL. The relativistic factor γ is given by the one-photon resonance condition:

$$E_3 - E_1 = \gamma(1+\beta)\hbar\omega_p. \tag{3.4}$$

The frequency of the Stokes x-ray laser can be then determined depending on the geometry of the setup. For copropagating pump and Stokes beams (implying a two-color XFEL), the photon energy of the Stokes laser is smaller than that of the pump laser since $E_2 > E_1$. The alternative that we put forward is to consider two crossed laser beams generated by a one-color SXFEL meeting the accelerated nuclei as shown schematically in Fig. 3.3b. The angle θ_S between the two beams is determined such

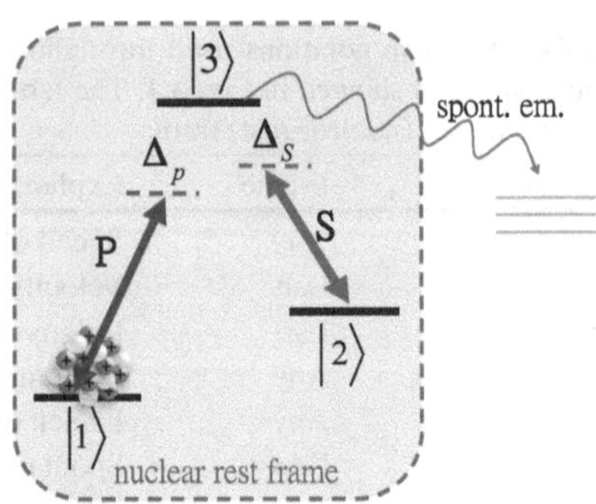

Fig. 3.4 The nuclear Λ-level scheme in the nuclear rest frame. The *blue arrow* illustrates the pump pulse, the *red arrow* depicts the Stokes pulse and all nuclear population is initially in $|1\rangle$ state

that in the nuclear rest frame the pump and Stokes photons fulfill the resonances with the two different nuclear transitions. The values of γ, $\hbar\omega_S$ and θ_S for NCPT for the nuclear systems under consideration are given in Table 3.2. The separation of the pump and Stokes beams out of the original XFEL beam requires dedicated x-ray optics such as the diamond mirrors [26–28] developed for the XFELO. X-ray reflections can also help tune the intensity of the two beams. The coherence between the two ground states is crucial for successful NCPT via STIRAP. Since in our case the lifetime of $|2\rangle$ is much longer than the laser pulse durations, decoherence is related to the unstable central frequencies and short coherence times of the pump and Stokes lasers. Our one-color XFEL crossed-beam setup accommodates the present lack of two-color x-ray coherent sources and reduces the effect of laser central frequency jumps to equal detunings in the pump and Stokes pulses. However, the crossed-beam setup suffers from a number of drawbacks related to spatial and temporal overlap of the laser and ion beams, as it will be addressed shortly in the following.

The XFEL-nuclei interaction in the nuclear rest frame is illustrated in Fig. 3.4. The nuclear dynamics is governed by the master equation for the nuclear density matrix $\widehat{\rho}(t)$ [8]

$$\frac{\partial}{\partial t}\widehat{\rho} = \frac{1}{i\hbar}\left[\widehat{H}, \widehat{\rho}\right] + \widehat{\rho}_s, \tag{3.5}$$

Table 3.2 Laser and nuclear beam parameters. The accelerated nuclei have the relativistic factor γ, determined by the one-photon resonance condition $\gamma(1+\beta)\hbar\omega_p = ck_{31}$. For the copropagating-beams setup, $\hbar\omega_S$ denotes the Stokes photon energy. The pump (copropagating beams) or both pump and Stokes lasers (crossed beams) photon energies are 12.4 keV for SXFEL and 25 keV for XFELO, respectively. For the crossed-beam setup, the angle θ_S between the pump and Stokes beams shown in Fig. 3.3b is given in rad

	SXFEL			XFELO		
Nucleus	γ	θ_S (rad)	$\hbar\omega_S$ (keV)	γ	θ_S (rad)	$\hbar\omega_S$ (keV)
^{185}Re	11.5	1.4544	6.93	5.7	1.4596	13.97
^{97}Tc	22.6	1.3836	7.36	11.2	1.3848	14.83
^{154}Gd	50.1	0.6407	11.17	24.8	0.6408	22.52
^{168}Er	72.0	0.4260	11.85	35.7	0.4260	23.88

with the interaction Hamiltonian

$$\widehat{H} = -\frac{\hbar}{2} \begin{pmatrix} 0 & 0 & \Omega_p^* \\ 0 & -2\left(\Delta_p - \Delta_S\right) & \Omega_S^* \\ \Omega_p & \Omega_S & 2\Delta_p \end{pmatrix}, \tag{3.6}$$

and the decoherence matrix

$$\widehat{\rho_s} = \frac{\Gamma}{2} \begin{pmatrix} 2B_{31}\rho_{33} & 0 & -\rho_{13} \\ 0 & 2B_{32}\rho_{33} & -\rho_{23} \\ -\rho_{31} & -\rho_{32} & -2\rho_{33} \end{pmatrix}. \tag{3.7}$$

The initial conditions are

$$\rho_{ij}(0) = \delta_{i1}\delta_{1j}. \tag{3.8}$$

The angle θ is zero for the pump laser and $\theta = \theta_S$ for the Stokes laser. So far, Eqs. 3.5–3.8 are the standard approaches for any quantum system like that showed in Fig. 3.4. The nuclear physics and relativitic treatment will enter the whole calculation when deriving the values of Rabi frequencies [22, 30] in the nuclear rest frame. The derivation of Rabi frequencies in the lab frame has been presented in Sect. 2.1. We merely quote the results here.

$$\begin{aligned} \Omega_p(t) &= \frac{1}{\hbar}\langle 3|\widehat{H}_I|1\rangle \\ &= \frac{4}{\hbar}\sqrt{\frac{\pi I_p^{\mathrm{eff}}(t)}{c\epsilon_0}}\sqrt{\frac{(2I_1+1)(L_{31}+1)}{L_{13}}}\frac{k_{31}^{L_{31}-1}}{(2L_{13}+1)!!}\sqrt{\mathbb{B}(\varepsilon/\mu\, L_{13})}, \end{aligned} \tag{3.9}$$

$$\begin{aligned} \Omega_S(t) &= \frac{1}{\hbar}\langle 3|\widehat{H}_I|2\rangle \\ &= \frac{4}{\hbar}\sqrt{\frac{\pi I_S^{\mathrm{eff}}(t)}{c\epsilon_0}}\sqrt{\frac{(2I_2+1)(L_{23}+1)}{L_{23}}}\frac{k_{32}^{L_{23}-1}}{(2L_{23}+1)!!}\sqrt{\mathbb{B}(\varepsilon/\mu\, L_{23})}, \end{aligned} \tag{3.10}$$

where $I_{p(S)}^{\mathrm{eff}}(t)$ is the Gaussian pump (Stokes) pulse in the nuclear rest frame:

$$I_{p(S)}^{\mathrm{eff}}(t) = \gamma^2\left(1+\beta\cos\theta\right)^2 I_{p(S)}^{\mathrm{eff}}\mathrm{Exp}\left\{-\left[\frac{\gamma(1+\beta\cos\theta)(t-\tau_{p(S)})}{T_{p(S)}}\right]^2\right\}. \tag{3.11}$$

The slowly varying effective Rabi frequencies $\Omega_{p(S)}(t)$ in the nuclear rest frame for nuclear transitions of electric (ε) or magnetic (μ) multipolarity L are given by [8, 22]

$$\Omega_{p(S)}(t) = \frac{4\sqrt{\pi}}{\hbar}\left[\frac{\gamma^2(1+\beta\cos\theta)^2 I^{\text{eff}}_{p(S)}(L_{1(2)3}+1)(2I_{1(2)}+1)\mathbb{B}(\varepsilon/\mu\, L_{1(2)3})}{c\epsilon_0 L_{1(2)3}}\right]^{1/2}$$
$$\times \frac{k_{31(2)}^{L_{1(2)3}-1}}{(2L_{1(2)3}+1)!!}\text{Exp}\left\{-\left[\frac{\gamma(1+\beta\cos\theta)(t-\tau_{p(S)})}{\sqrt{2}T_{p(S)}}\right]^2\right\}. \tag{3.12}$$

Here we have expressed the nuclear multipole moment with the help of the reduced transition probabilities $\mathbb{B}(\varepsilon/\mu\, L)$ introduced in Sect. 2.1 following the approach developed in Sect. 2.1 and Ref. [22]. This allows for a unified treatment of the laser-nucleus interaction for both dipole-allowed ($E1$) and dipole-forbidden nuclear transitions. All the laser quantities have been transformed in Eq. (3.12) into the nuclear rest frame, leading to

- the angular frequency $\gamma(1+\beta\cos\theta)\omega_{p(S)}$,
- bandwidth $\gamma(1+\beta\cos\theta)\Gamma_{p(S)}$,
- pulse duration $T_{p(S)}/(\gamma(1+\beta\cos\theta))$,
- laser peak intensity $\gamma^2(1+\beta\cos\theta)^2 I_{p(S)}$.

Let us turn to the time scale analysis in the nuclear rest frame, with focus on comparing the XFEL pulse duration with the lifetimes of the nuclear excited states. Typically, the Λ-level scheme in Fig. 3.4 is not closed, i.e. the population in $|3\rangle$ will not only decay to $|1\rangle$ and $|2\rangle$ but also to other low energy levels through spontaneous γ-decay or by other decay mechanisms such as α decay. This open feature of $|3\rangle$ speaks against direct pumping. On the other hand, the delivered XFEL pulse duration is fixed due to the designs. Comparing the lifetime of the state $|3\rangle$ with the XFEL pulse duration allows us to identify two situations:

(i) $1/\Gamma \geq \frac{T_{p(S)}}{\gamma(1+\beta\cos\theta)}$, i.e., the lifetime of $|3\rangle$ is longer than the pulse duration. Since the nucleus can stay in $|3\rangle$ long enough, apart from STIRAP, also NCPT via sequential isolated pulses such as π pulses, i.e., pulses that transfer the complete nuclear state population from one state to another, is possible. A first π pulse can pump the nuclei from $|1\rangle$ to $|3\rangle$, followed by a second Stokes π pulse that drives the $|3\rangle \rightarrow |2\rangle$ decay. The latter scenario lacks the robustness of STIRAP, having a sensitive dependence on the laser intensities.

(ii) $1/\Gamma < \frac{T_{p(S)}}{\gamma(1+\beta\cos\theta)}$, i.e., the lifetime of $|3\rangle$ is shorter than the pulse duration. Because of the high decay rate of $|3\rangle$, separated single pulses cannot produce NCPT and STIRAP provides the only possibility for population transfer.

In general, situation (i) is related to nuclear excitations of tens up to hundreds of keV, such that $\gamma \sim 10$. These low-lying levels have however energy widths of about 1 μeV or less, orders of magnitude smaller than the photon energy spread. In this case only a fraction of the incoming photons will drive the nuclear transition, leading to a small effective intensity [22]. For case (ii), the required γ for driving MeV transitions is on the order of 20–100. Typically, such transitions have widths ($\sim$1 eV) larger than the bandwidth of the XFELO or SXFEL. The effective and nominal laser intensity have in this case the same value, an advantage of the high-γ regime. Moreover, for narrow-width excitations (i) it is necessary to first find the laser bandwidth window of the nuclear transition, since most of the transition energy values are not known with such precision. Once found, our procedure of considering an effective intensity which is scaled according to the number of resonant photons should provide the correct approach for a zero-detuning situation. For the case (ii) where the MeV nuclear transitions have eV widths, it is only necessary to tune the laser photons in the corresponding

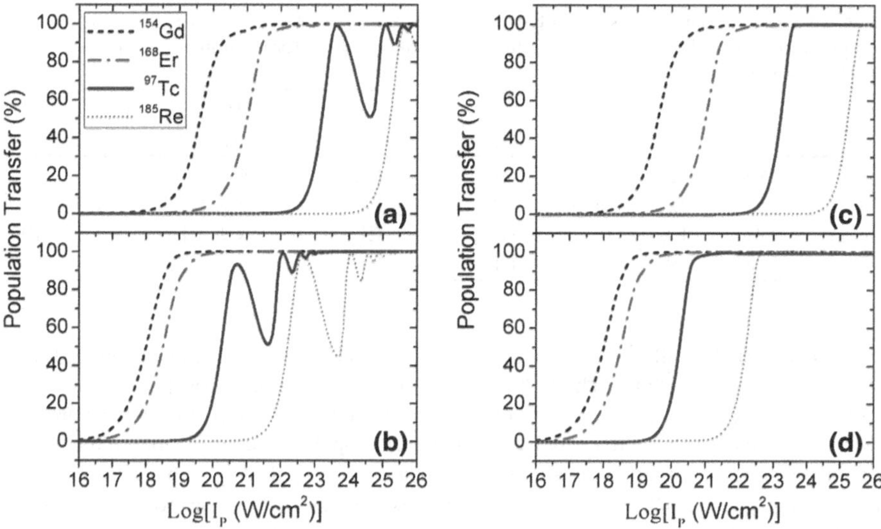

Fig. 3.5 NCPT for several nuclei as a function of the pump XFEL peak intensity using SXFEL (**a, c**) and XFELO (**b, d**) parameters. For the crossed-beams setup (a) and (b), the Stokes laser intensities were chosen $I_S = 0.02I_p$ for ^{185}Re, $I_S = 0.34I_p$ for ^{168}Er, $I_S = 0.81I_p$ for ^{154}Gd and $I_S = 20.82I_p$ for ^{97}Tc, respectively, according to the π pulse intensity ratios I_S^{π}/I_p^{π}. In the two-color setup (c) and (d), $I_S = 0.03I_p$ for ^{185}Re, $I_S = 0.35I_p$ for ^{168}Er, $I_S = 0.90I_p$ for ^{154}Gd and $I_S = 35.06I_p$ for ^{97}Tc. All detunings are $\Delta_p = \Delta_S = 0$. See discussion in the text and Tables 3.2 and 3.3 for further parameters

energy window. A list of parameters for a number of nuclei with suitable transitions for both (i) and (ii) regimes is presented in Tables 3.2 and 3.3.

3.2.2 Results and Discussion

In Fig. 3.5 we compare our calculated population transfer for several cases in both regimes (i) and (ii) using SXFEL (Fig. 3.5a) and XFELO (Fig. 3.5b) parameters in a crossed-beam single-color XFEL setup. For the two-color copropagating beams setup, the results using SXFEL and XFELO parameters are showed in Fig. 3.5c and d, respectively. We investigate first the efficiency of NCPT for nuclear three-level systems that do not present a metastable state. The considered nuclear transition energies, multipolarities and reduced matrix elements are given in Table 3.3. The choice of nuclei is related to nuclear data availability and the lifetime values of state $|3\rangle$ required by the two parameter regimes (i) and (ii). The optimal set of laser parameters is obtained by a careful analysis of the dependence between pump peak intensity I_P and pulse delay $\tau_P - \tau_S$. A negative time delay corresponds to the π-pulse population transfer regime, while a positive one stands for STIRAP. For each value of I_P, the $\tau_P - \tau_S$ is chosen such that the NCPT reaches its maximum value. Two examples of this optimization process are demonstrated in Fig. 3.6. Each red dashed line on the contour plot illustrates the optimal cases, based on which we select the necessary laser intensities, showed in Fig. 3.5a.

Table 3.3 Nuclear parameters. E_i is the energy of state $|i\rangle$ with $i \in \{1, 2, 3\}$ (in keV) [31]. The multipolarities and reduced matrix elements (in Weisskopf units) for the transitions $|j\rangle \rightarrow |3\rangle$ with $j \in \{1, 2\}$ are also given

Nucleus	E_3	E_2	E_1	$\epsilon/\mu L$		$\mathbb{B}(\varepsilon/\mu\, L)$ (wsu)	
				$\lvert 1\rangle \rightarrow \lvert 3\rangle$	$\lvert 2\rangle \rightarrow \lvert 3\rangle$	$\lvert 1\rangle \rightarrow \lvert 3\rangle$	$\lvert 2\rangle \rightarrow \lvert 3\rangle$
^{185}Re	284	125	0	E2	M1	6.4×10	3.7×10^{-1}
^{97}Tc	657	324	96.57	E2	E1	5×10^{2}	6.7×10^{-5}
^{154}Gd	1241	123	0	E1	E1	4.4×10^{-2}	4.9×10^{-2}
^{168}Er	1786	79	0	E1	E1	3.2×10^{-3}	9.1×10^{-3}

For regime (i) that allows NCPT via both π pulses and STIRAP, we considered the lowest three nuclear levels of ^{185}Re. The γ, θ_S (for one-color setup), and Stokes photon energy E_S (for two-color scheme) are listed in Table 3.2. In the one-color setup, NCPT is achieved at lower intensities via sequential π pulses. At the exact π-pulse value of the pump intensity, a peak in the nuclear population transfer for ^{185}Re can be observed, at $I_p = 6 \times 10^{25}$ W/cm^2 in Fig. 3.5a and $I_p = 6 \times 10^{22}$ W/cm^2 in Fig. 3.5b. With increasing I_p in the crossed-beam setup (Fig. 3.5a, b), the ^{185}Re nuclei are only partially excited to state $|2\rangle$ and the NCPT yield starts to oscillate. The amplitude and frequency of the oscillations are varying as a result of our pulse delay optimization procedure. At sufficient intensities in the pulse overlap regime STIRAP becomes preferable as compared to the π pulses mechanism due to the lack of oscillations. The plateau at 100 % population transfer indicates that NCPT via STIRAP alone is reached. In the two-color copropagating beams scheme (Fig. 3.5c, d), the pulse shape of pump and that of Stokes are the same in the nuclear rest frame. This makes STIRAP more efficient and thus preferable compared to the single-color setup, as the STIRAP plateau can be reached with lower laser intensities.

For case (ii), we present our results for ^{154}Gd and ^{168}Er, that require stronger nuclear acceleration with γ factors between 24 and 72 and fs pulse delays. The ^{154}Gd ground state population starts to be coherently channeled at about $I_p = 10^{17}$ W/cm^2 using XFELO and $I_p = 10^{19}$ W/cm^2 using SXFEL parameters, respectively. Up to $I_p = 10^{19}$ W/cm^2 (XFELO) and $I_p = 10^{21}$ W/cm^2 (SXFEL), more than 95 % of the nuclei reach state $|2\rangle$. In this case π pulses cannot provide the desired NCPT due to the fast spontaneous decay of state $|3\rangle$ in neither copropagating- nor cross-beam setups. The calculated intensities necessary for complete NCPT are within the designed intensities of the XFEL sources. Considering the operating and designed peak power of 20–100 GW [2–4, 6, 24] for SXFEL (and about three orders of magnitude less for XFELO) and the admirable focus achieved for x-rays of 7 nm [32], intensities could reach as high as 10^{17}–10^{18} W/cm^2 for XFELO [23] and 10^{21}–10^{22} W/cm^2 for SXFEL [2, 6]. As a recent result development, focusing the 10 keV photon beam with the reflective optics at SACLA is expected to be applicable for the generation of a nm-size hard x-ray laser [33]. This progress may render possible an XFEL intensity larger than 10^{22} W/cm^2.

One of the most relevant applications of NCPT is isomer pumping or depletion. In Fig. 3.5 we present our results for NCPT in ^{97}Tc nuclei starting from the $E_1 = 96.57$ keV isomeric state which has a half life of $\tau_1 = 91$ d. Like ^{185}Re, ^{97}Tc belongs to regime (i) such that NCPT at lower intensities can be achieved via π pulses in the crossed-beam setup. The intensity for which complete isomer depletion is achieved using SXFEL is $I_p = 4 \times 10^{23}$ W/cm^2. Due to the longer pulse duration of the XFELO and consequently higher losses via spontaneous decay of state $|3\rangle$, the peak population transfer at $I_p = 5.2 \times 10^{20}$ W/cm^2 reaches only 93 %

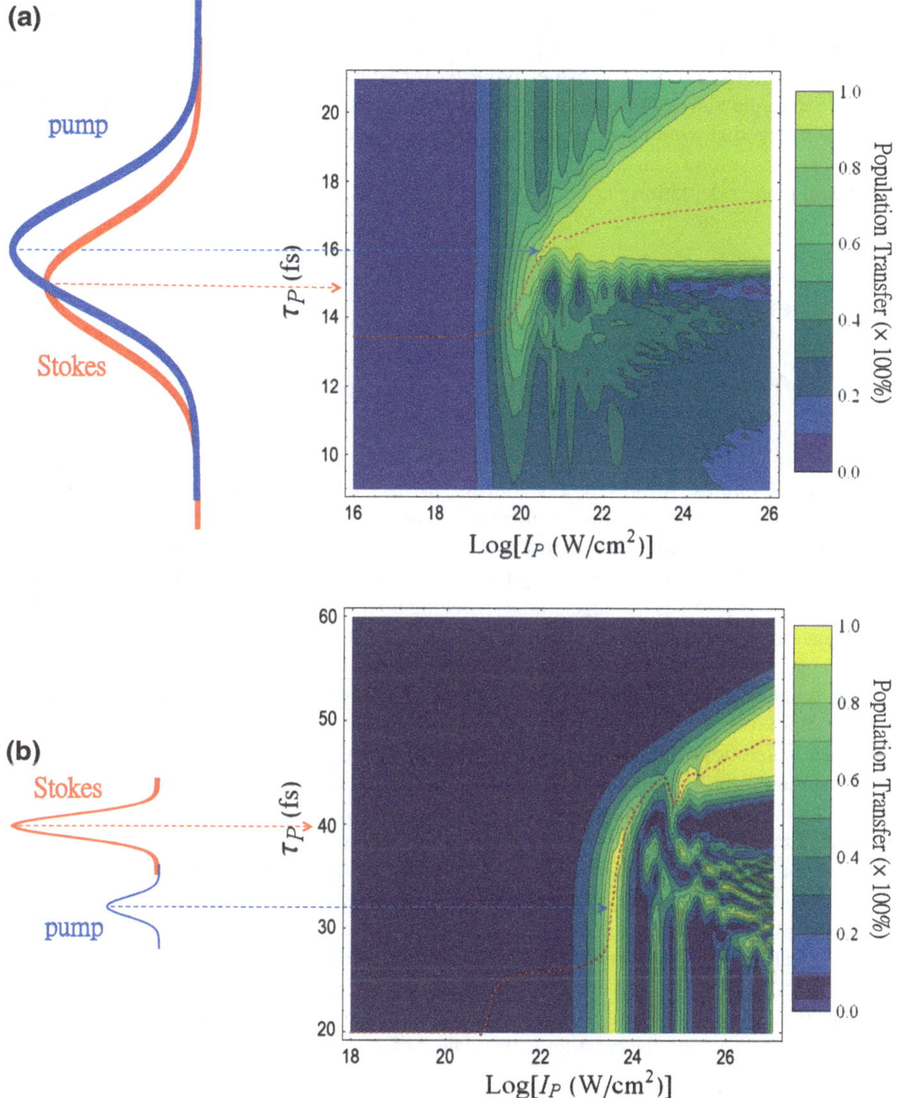

Fig. 3.6 The laser peak-intensity & pulse-delay dependent NCPT. **a** for ^{154}Gd, the Stokes peak position τ_S is fixed at 15 fs, and $I_S/I_p = 0.81$. Eq. (3.5) is numerically solved for $9 \leq \tau_p \leq 21$ fs and $10^{16} \leq I_p \leq 10^{26}$ W/cm^2. **b** for ^{97}Tc, the Stokes peak position τ_S is fixed at 40 fs, and $I_S/I_p = 20.82$. The numerical solution of Eq. (3.5) is displayed in the interval of $20 \leq \tau_p \leq 60$ fs and $10^{18} \leq I_p \leq 10^{27}$ W/cm^2. The NCPT curve of ^{154}Gd and ^{97}Tc in Fig. 3.5a are selected along the optimized *red dashed lines* in **a** and **b**, respectively

in Fig. 3.5b in the crossed-beam setup. For the copropagating beams setup, 100 % NCPT is achieved for the same intensity $I_p = 5.2 \times 10^{20}$ W/cm^2. Compared to the case of high-energy nuclear transitions (ii), the intensities required for isomer depletion are in this case larger, mainly due to the narrow transition width of state $|3\rangle$. Typically, triggering levels high above isomeric states, that would present the advantage of larger linewidths, are less well known. A detailed analysis of nuclear data in the search for the best candidate is therefore required for successful isomer depletion.

3.3 Experimental Facilities

3.3.1 Infrastructure and Table-Top Solutions

X-ray coherent light sources, listed in Table 3.4, are not available today at the few large ion acceleration facilities. At present a new materials research center MaRIE (Matter-Radiation Interactions in Extreme) providing both a fully coherent XFEL with photon energy of 50 keV and accelerated charged-particle beams is envisaged in the USA [34]. In addition, the photonuclear physics pillar of the Extreme Light Infrastructure (ELI) can provide simultaneously a compact XFEL as well as ion acceleration reaching up to 4–5 GeV [35]. At ELI, the combination of gamma-rays and acceleration of the nuclear target are already under consideration for nuclear resonance fluorescence experiments [35]. Furthermore, ELI is also envisaged to deliver gamma rays with energies of few MeV [35], which could be used for direct photoexcitation of giant dipole resonances [36].

Tabletop solutions for both ion acceleration and x-ray coherent light, showed in Table 3.5, would facilitate the experimental realization of isomer depletion in NCPT and nuclear batteries. Tabletop x-ray undulator sources are already operational [37], with a number of ideas envisaging compact x-ray FELs [38, 39]. Rapid progress spanning five orders of magnitude increase in the achieved light brightness within only two years has been reported [40–42]. In conjunction with the crystal cavities designed for the XFELO, such table-top devices have the potential to become a key tool for the release on demand of energy stored in nuclei at large ion accelerator facilities. Alternatively, the exciting forecast of compact shaped-foil-target ion accelerators [43, 44], foil-and-gas target [45] and radiation pressure acceleration [46–50] together with microlens beam focusing [51] are likely to provide a viable table-top solution to be used together with the existing large-scale XFELs.

3.3.2 The Influence of $\Delta\gamma$ and $\Delta\theta_S$ for the One-Color Setup

NCPT via STIRAP is sensitive to the fulfillment of the two-photon resonance condition $\Delta_p = \Delta_S$. This involves on the one hand precise knowledge of the nuclear transition energy and on the other hand good control of the laser frequency and therefore of the nuclear acceleration. The former is usually attained in nuclear forward scattering by scanning first for the position of the nuclear resonance. As for the latter, in our setup, the relativistic factor γ influences the detunings and the effective pump and Stokes intensities and Rabi frequencies.

Table 3.4 Specifications of the existent and forthcoming XFEL facilities

Facility	LCLS	SACLA	European XFEL	MaRIE	XFELO
Ref.	[3, 52, 53]	[5, 33]	[4]	[34]	[23]
Location (Country)	SLAC (USA)	Spring-8 (Japan)	DESY (Germany)	LANL (USA)	ANL (USA)
Status	Operation	Operation	Construction	Proposal	Proposal
Photon energy (keV)	5–25	<15.5	0.25–12.4	50	5–20
Coherence time (fs)	0.55–2	Not clear	0.2–1.4	Not clear	1000
Pulse duration (fs)	10–300	<100	100	<100	1000
Peak power (GW)	∼10	5–29	20–150	10	0.001
Photon per pulse	10^{11}–10^{12}	5×10^{11}	10^{12}–10^{14}	10^{11}	10^{9}
Beam size (FWHM) (μm^2)	1.3 × 1.3–3000 × 3000	0.95 × 1.2–33 × 33	55 × 55–90 × 90	Not clear	Not clear

Table 3.5 Specifications of some table-top x-ray sources

Scheme	Plasma wiggler	Thomson back scattering	Magnet undulator	HHG (AMO)	Carbon nanotube
Ref.	[40–42, 54]	[55, 56]	[37–39, 57]	[58]	[59]
Photon energy (keV)	1–150	0.4–1000	0.14	>0.5	50–500
Coherent	Spatially	Not	Spatially	Spatially	Not
Peak brilliance	10^{21}–10^{23}	3×10^{17}	1.3×10^{17}	6×10^{7} photons/s	Not clear
Pulse duration (fs)	10–30	Not clear	10	0.01	Not clear

So far, as showed in Fig. 3.7a we consider an ideal case, using a monoenergetic beam from ion accelerators to bridge the gap between nuclear transition and x-ray laser energies. The designed γ and θ_S are important parameters for achieving NCPT. In the low γ region, the forthcoming FAIR at GSI will provide high quality ion beams with energies up to 45 GeV/u [60]. The corresponding γ limit is about 48 and the precision $\Delta E/E \sim 2\times10^{-4}$. For the high γ region, the Large Hadron Collider (LHC) is currently the only suitable ion accelerator which can accelerate $^{208}Pb^{82+}$ up to $\gamma = 2963.5$ with low energy spread of about 10^{-4} [61]. LHC can also accelerate lighter ions to energies larger than 100 GeV [62]. For the strong acceleration regime, the resonance condition corresponds to an energy spread of the ion beam of 10^{-5}. This issue becomes more problematic for NCPT of nuclei in the moderate acceleration regime where the resonance condition requires a more precise γ value, $\Delta\gamma/\gamma = 10^{-6}$. On the other hand, the European XFEL will deliver laser pulses with the divergence angle of about 10^{-6} rad [4]. This causes the missmatch of $\Delta_p \neq \Delta_S$ together with the energy spread ΔE of an ion beam. To address the realistic case in Fig. 3.7b, we numerically solve Eq. (3.5) with $\gamma \to \gamma + \Delta\gamma$ and $\theta_s \to \theta_s + \Delta\theta_s$. Our results are presented in Fig. 3.8, where the two errors

$\Delta\gamma$ and $\Delta\theta_s$ are scanned. We find NCPT maintains values of around 80 % in the region of $\theta_S \pm 10^{-5}$ rad and $\Delta\gamma/\gamma = \pm 10^{-6}$ for ^{154}Gd and ^{168}Er.

In the following, we attempt to connect the two-photon resonance condition $\Delta_p = \Delta_S$ to $\Delta\gamma$ and $\Delta\theta_S$ in the one-color setup illustrated in Fig. 3.7b. Let us begin with the one photon detunings in the nuclear rest frame

$$\begin{aligned}\Delta_P &= \omega_P\gamma(1+\beta) - ck_{31},\\ \Delta_S &= \omega_S\gamma(1+\beta\cos\theta_S) - ck_{32}.\end{aligned} \tag{3.13}$$

Substituting

$$\begin{aligned}\beta &\rightarrow \sqrt{1-\frac{1}{\gamma^2}},\\ \gamma &\rightarrow \gamma+\Delta\gamma,\\ \theta_S &\rightarrow \theta_S+\Delta\theta_S,\end{aligned} \tag{3.14}$$

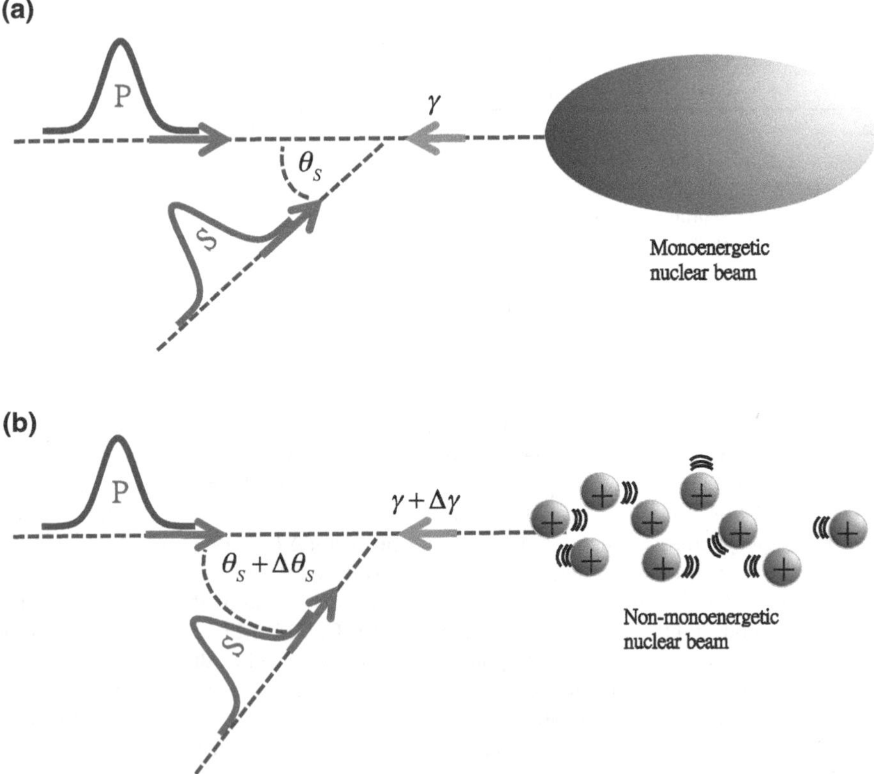

Fig. 3.7 **a** The ideal case. All nuclei fly with the same velocity of the designed γ, and all the XFEL photons propagate along two directions at the angle θ_S. **b** The real case. The nuclear beam is not monoenergenic, and the x-ray photons propagate along two directions at the angle $\theta_S + \Delta\theta_S$

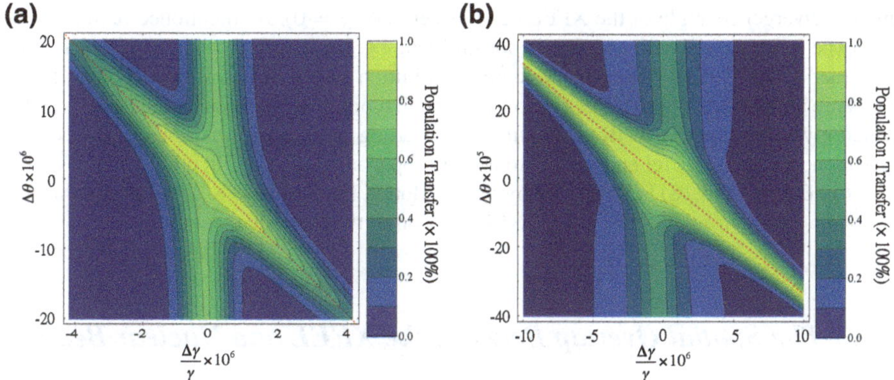

Fig. 3.8 $\Delta\theta_S-$ and $\Delta\gamma$-dependent NCPT in the one-color scheme. **a** For ^{168}Er, $I_p = 2.85 \times 10^{22}$ W/cm^2 and $I_S = 9.57 \times 10^{21}$ W/cm^2. **b** For ^{154}Gd, $I_p = 6.78 \times 10^{21}$ W/cm^2 and $I_S = 5.49 \times 10^{21}$ W/cm^2. The color coding shows the different percentage of population transfer with SXFEL. The *red dashed line* depicts Eq. (3.18)

into $\Delta_p = \Delta_S$, then we obtain

$$\omega_p\,(\gamma + \Delta\gamma)\left[1 + \sqrt{1 - \frac{1}{(\gamma + \Delta\gamma)^2}}\right] - ck_{31}$$
$$= \omega_S\,(\gamma + \Delta\gamma)\left[1 + \sqrt{1 - \frac{1}{(\gamma + \Delta\gamma)^2}}\cos(\theta_S + \Delta\theta_S)\right] - ck_{32}. \tag{3.15}$$

Using Taylor's expansion, and treating $\Delta\gamma$ and $\Delta\theta_S$ as small variables, the expansion becomes

$$\omega_p\gamma\left(1 + \sqrt{1 - \frac{1}{\gamma^2}}\right) + \frac{\omega_p\Delta\gamma\left(1 + \sqrt{1 - \frac{1}{\gamma^2}}\right)}{\sqrt{1 - \frac{1}{\gamma^2}}} + \ldots - ck_{31}$$
$$= \omega_S\gamma\left(1 + \cos\theta_S\sqrt{1 - \frac{1}{\gamma^2}}\right) - \omega_S\gamma\Delta\theta_S\sin\theta_S\sqrt{1 - \frac{1}{\gamma^2}} + \ldots - ck_{32}. \tag{3.16}$$

Finally, we get the ideal 0th order two-photon resonance condition

$$\omega_p\gamma\,(1 + \beta) - ck_{31} = \omega_S\gamma\,(1 + \beta\cos\theta_S) - ck_{32}, \tag{3.17}$$

and the 1st order two-photon resonance condition (using $\omega_S = \omega_p$ for one-color setup)

$$\Delta\theta_S = -\frac{1 + \beta}{\beta^2\sin\theta_S}\left(\frac{\Delta\gamma}{\gamma}\right). \tag{3.18}$$

Equation (3.17) is the condition for implementing STIRAP in an ideal case, i.e., the kinetic energy distribution of the nuclear beam is perfectly monoenergetic at the designed $\gamma(\Delta\gamma = 0)$

and the divergence angle of the XFEL beam is zero ($\Delta\theta_S = 0$). As mentioned before, in the real experiments the ideal condition is not fulfilled, e.g. the divergence angle of XFEL is on the order of 10^{-6} rad [4], and the velocity distribution of ion beams from LHC and future FAIR are not perfect. Then one has to consider the first order two-photon resonance Eq. (3.18) which gives a less strict STIRAP requirement, leading to an additional match between the non-monoenergetic nuclei and the photons that propagate along the direction at angles other than θ_S. Equation (3.18) is illustrated by the red dotted line in Fig. 3.8, and the agreement is verified by comparing it with the high NCPT region of the numerical solution.

3.3.3 The Spatial Overlap Between the XFEL and Nuclear Beams

A further study of the overlap efficiency for the laser beams and ion bunches shows that the copropagating laser beams setup is more advantageous. In Fig. 3.9, using the LHC beam size parameters [61] and a 10 μm focusing of the XFEL beam, we estimate that for (a) copropagating laser beams up to 10^5 nuclei meet the laser focus per bunch and laser pulse, while for (b) crossed laser beams this number reduces to 30. The extreme temporal and spatial fine-tuning required to match the overlaps of a bunched ion beam with the two laser beams in the crossed-beam setup is however at present out of reach. A continuous ion beam, on the other

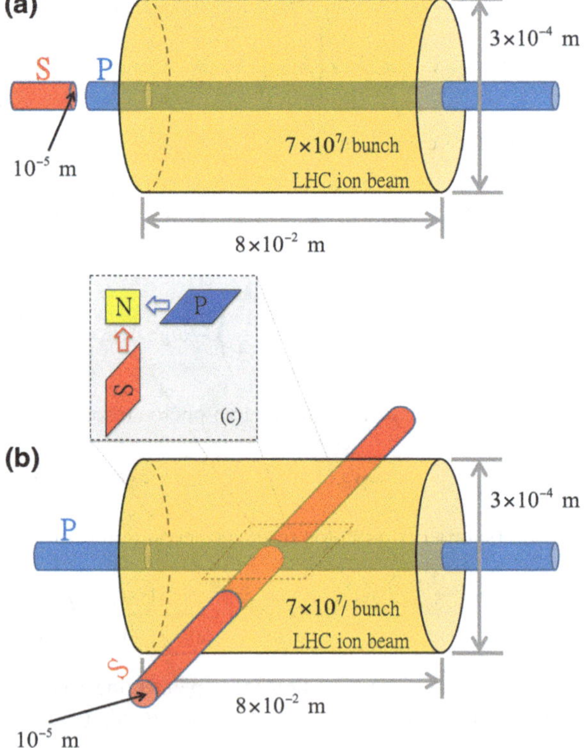

Fig. 3.9 The spatial overlap between the nuclear beam, pump pulse and the Stokes pulse in the nuclear rest frame for (**a**) the two-*color* scheme and (**b**) the one-*color* scheme. The parameters of the LHC ion beam are considered [61]. The *blue* (*red*) 'pipeline' is the volume that the pump (Stokes) XFEL pulse flies through, and the light orange *cylinder* denotes the nuclear bunch. **c** the required XFEL wave front for implementing STIRAP in a one-*color* setup. The *blue* (*red*) parallelogram depicts the wave front of pump (Stokes) pulse and the *yellow square* illustrates the nuclear area

hand, has the disadvantage of much lower ion density at the overlap with the pump and Stokes beam. Furthermore, the necessary time delay between pump and Stokes and the adiabaticity condition for STIRAP will be in this case only fulfilled for ions at the diagonal line of the overlap area. In order to maintain the pulse delay and the adiabaticity condition for the whole overlap region with the nuclear beam, a special laser pulse front as presented in Fig. 3.9c is required. We conclude therefore that for a number of technical and conceptual reasons, the two-color copropagating beams scheme has better chances to be realized experimentally in the near future.

3.4 Summary

In summary, we have investigated the NCPT using a collider system composed of a fully coherent XFEL together with an ion accelerator, and considered the interaction between an accelerated nuclear bunch and two XFEL pulses. This system is showed to be a powerful tool for studying the radiation-nuclei interaction such that one can excite a high energy nuclear transition with relatively low energy hard x-ray photons via Doppler-blue-shift. Two schemes, the two-color-linear and one-color-cross geometry as showed in Fig. 3.3, have been proposed. The required parameters of the used two laser pulses and the nuclear bunch were derived for achieving a complete NCPT between two nuclear ground levels directly (indirectly) via a third level using two π-pulses (STIRAP) method. We have selected the necessary laser peak intensities with an optimization process, scanning both the laser peak intensity and the time delay between two pulses for NCPT. An XFELO (SXFEL) laser peak intensity of around $10^{18}-10^{19}$ W/cm^2 ($10^{20}-10^{21}$ W/cm^2) was found to be sufficient to be used to achieve 100 % NCPT for the ^{154}Gd and ^{168}Er nuclei. Also, the coherent ^{97}Tc isomer triggering was considered, and the required XFELO and SXFEL peak intensity found are around 10^{21} W/cm^2 and 3×10^{23} W/cm^2, respectively. Moreover, we have derived the first order two-photon resonance condition to connect the error of the XFEL divergence angle and that of a nuclear bunch. This additional condition gives a less strict requirement for the experimental implementation, and can be used to design the parameters of the laser and the nuclear bunches. For getting the number of the coherently excited nuclei via NCPT, we have considered the spatial overlap among two XFEL pulses and the LHC nuclear bunch. By using a two-color (one-color) scheme, up to 10^5 nuclei (30 nuclei) will be coherently excited. Consequently, the two-color scheme is found to be much more efficient for NCPT.

References

1. J. Feldhaus, E. Saldin, J. Schneider, E. Schneidmiller, M. Yurkov, Possible application of x-ray optical elements for reducing the spectral bandwidth of an x-ray sase fel. Opt. Commun. **140**, 341 (1997)
2. E. Saldin, E. Schneidmiller, Y. Shvyd'ko, M. Yurkov, X-ray fel with a mev bandwidth. Nucl. Instrum. Methods Phys. Res. Sect A **475**, 357 (2001)
3. J. Arthur et al., Linac Coherent Light Source (LCLS), Conceptual Design Report (SLAC, Stanford, CA, 2002)
4. M. Altarelli et al., XFEL: The European X-Ray Free-Electron Laser, Technical Design Report (DESY, Hamburg, 2009)

5. M. Yabashi, T. Ishikawa, *XFEL/SPring-8 Beamline Technical Design Report Ver. 2.0* (RIKEN-JASRI XFEL Project Head, Office, 2010).
6. Linac coherent light source lcls-ii, slac, stanford, https://slacportal.slac.stanford.edu/sites/lcls_public/lcls_ii
7. A.Y. Dzyublik, Triggering of nuclear isomers by x-ray laser. JETP Lett. **92**, 130 (2010)
8. K. Bergmann, H. Theuer, B.W. Shore, Coherent population transfer among quantum states of atoms and molecules. Rev. Mod. Phys. **70**, 1003 (1998)
9. T.J. Bürvenich, J. Evers, C.H. Keitel, Nuclear quantum optics with x-ray laser pulses. Phys. Rev. Lett. **96**, 142501 (2006)
10. P. Walker, G. Dracoulis, Energy traps in atomic nuclei. Nature **399**, 35 (1999)
11. K.W.D. Ledingham, P. McKenna, R.P. Singhal, Applications for nuclear phenomena generated by ultra-intense lasers. Science **300**, 1107 (2003)
12. A. Pálffy, J. Evers, C.H. Keitel, Isomer triggering via nuclear excitation by electron capture. Phys. Rev. Lett. **99**, 172502 (2007)
13. C.B. Collins, C.D. Eberhard, J.W. Glesener, J.A. Anderson, Depopulation of the isomeric state $^{180}\mathrm{Ta}^m$ by the reaction $^{180}\mathrm{Ta}^m(\gamma,\gamma')\,^{180}\mathrm{Ta}$. Phys. Rev. C **37**, 2267 (1988)
14. D. Belic, C. Arlandini, J. Besserer, J. de Boer, J.J. Carroll, J. Enders, T. Hartmann, F. Käppeler, H. Kaiser, U. Kneissl, M. Loewe, H.J. Maier, H. Maser, P. Mohr, P. von Neumann-Cosel, A. Nord, H.H. Pitz, A. Richter, M. Schumann, S. Volz, A. Zilges, Photoactivation of $^{180}\mathrm{Ta}^m$ and its implications for the nucleosynthesis of nature's rarest naturally occurring isotope. Phys. Rev. Lett. **83**, 5242 (1999)
15. J. Carroll, An experimental perspective on triggered gamma emission from nuclear isomers. Laser Phys. Lett. **1**, 275 (2004)
16. C.B. Collins, F. Davanloo, M.C. Iosif, R. Dussart, J.M. Hicks, S.A. Karamian, C.A. Ur, I.I. Popescu, V.I. Kirischuk, J.J. Carroll, H.E. Roberts, P. McDaniel, C.E. Crist, Accelerated emission of gamma rays from the 31-yr isomer of $^{178}\mathrm{Hf}$ induced by x-ray irradiation. Phys. Rev. Lett. **82**, 695 (1999)
17. D.P. McNabb, J.D. Anderson, J.A. Becker, M.S. Weiss, Comment on "accelerated emission of gamma rays from the 31-yr isomer of $^{178}\mathrm{Hf}$ induced by x-ray irradiation". Phys. Rev. Lett. **84**, 2542 (2000)
18. P. von Neumann-Cosel, A. Richter, Comment on "accelerated emission of gamma rays from the 31-yr isomer of $^{178}\mathrm{Hf}$ induced by x-ray irradiation". Phys. Rev. Lett. **84**, 2543 (2000)
19. S. Olariu, A. Olariu, Comment on "accelerated emission of gamma rays from the 31-yr isomer of $^{178}\mathrm{Hf}$ induced by x-ray irradiation". Phys. Rev. Lett. **84**, 2541 (2000)
20. C.B. Collins, F. Davanloo, M.C. Iosif, R. Dussart, J.M. Hicks, S.A. Karamian, C.A. Ur, I.I. Popescu, V.I. Kirischuk, H.E. Roberts, P. McDaniel, C.E. Crist, Collins et al. reply. Phys. Rev. Lett. **84**, 2544 (2000)
21. M.O. Scully, M.S. Zubairy, *Quantum Optics* (Cambridge University Press, Cambridge, 1997)
22. A. Pálffy, J. Evers, C.H. Keitel, Electric-dipole-forbidden nuclear transitions driven by super-intense laser fields. Phys. Rev. C **77**, 044602 (2008)
23. K.-J. Kim, Y. Shvyd'ko, S. Reiche, A proposal for an x-ray free-electron laser oscillator with an energy-recovery linac. Phys. Rev. Lett. **100**, 244802 (2008)
24. P. Emma, R. Akre, J. Arthur, R. Bionta, C. Bostedt, J. Bozek, A. Brachmann, P. Bucksbaum, R. Coffee, F. Decker et al., First lasing and operation of an ångstrom-wavelength free-electron laser. Nat. Photonics **4**, 641 (2010)
25. J. Amann, W. Berg, V. Blank, F. Decker, Y. Ding, P. Emma, Y. Feng, J. Frisch, D. Fritz, J. Hastings et al., Demonstration of self-seeding in a hard-x-ray free-electron laser. Nat. Photonics **6**, 693 (2012)
26. Y. Shvyd'ko, S. Stoupin, A. Cunsolo, A. Said, X. Huang, High-reflectivity high-resolution x-ray crystal optics with diamonds. Nat. Phys. **6**, 196 (2010)
27. Y. Shvyd'Ko, S. Stoupin, V. Blank, S. Terentyev, Near-100% bragg reflectivity of x-rays. Nat. Photon. **5**, 539 (2011)
28. R.R. Lindberg, K.-J. Kim, Y. Shvyd'ko, W.M. Fawley, Performance of the x-ray free-electron laser oscillator with crystal cavity. Phys. Rev. ST Accel. Beams **14**, 010701 (2011)

29. M.O. Scully, E.S. Fry, C.H.R. Ooi, K. Wódkiewicz, Directed spontaneous emission from an extended ensemble of n atoms: Timing is everything. Phys. Rev. Lett. **96**, 010501 (2006)
30. J.M. Blatt, V.F. Weisskopf, in *Theoretical Nuclear Physics* (Courier Dover Publications, New York, 1991)
31. Nuclear structure and decay databases, http://www.nndc.bnl.gov
32. H. Mimura, S. Handa, T. Kimura, H. Yumoto, D. Yamakawa, H. Yokoyama, S. Matsuyama, K. Inagaki, K. Yamamura, Y. Sano et al., Breaking the 10 nm barrier in hard-x-ray focusing. Nat. Phys. **6**, 122 (2009)
33. H. Yumoto, H. Mimura, T. Koyama, S. Matsuyama, K. Tono, T. Togashi, Y. Inubushi, T. Sato, T. Tanaka, T. Kimura et al., Focusing of x-ray free-electron laser pulses with reflective optics. Nat. Photonics **7**, 43 (2012)
34. Marie: Matter-radiation interactions in extremes experimental facility, http://marie.lanl.gov/
35. Extreme light infrastructure, http://www.extreme-light-infrastructure.eu/reports.php
36. H.A. Weidenmüller, Nuclear excitation by a zeptosecond multi-mev laser pulse. Phys. Rev. Lett. **106**, 122502 (2011)
37. M. Fuchs, R. Weingartner, A. Popp, Z. Major, S. Becker, J. Osterhoff, I. Cortrie, B. Zeitler, R. Horlein, G.D. Tsakiris, U. Schramm, T.P. Rowlands-Rees, S.M. Hooker, D. Habs, F. Krausz, S. Karsch, F. Gruner, Laser-driven soft-x-ray undulator source. Nat. Phys. **5**, 826 (2009)
38. K. Nakajima, Compact x-ray sources: Towards a table-top free-electron laser. Nat. phys. **4**, 92 (2008)
39. F. Grüner, S. Becker, U. Schramm, T. Eichner, M. Fuchs, R. Weingartner, D. Habs, J. Meyer-ter Vehn, M. Geissler, M. Ferrario, L. Serafini, B. van der Geer, H. Backe, W. Lauth, S. Reiche, Design considerations for table-top, laser-based vuv and x-ray free electron lasers. Appl. Phys. B **86**, 431 (2007)
40. S. Kneip, S.R. Nagel, C. Bellei, N. Bourgeois, A.E. Dangor, A. Gopal, R. Heathcote, S.P.D. Mangles, J.R. Marquès, A. Maksimchuk, P.M. Nilson, K.T. Phuoc, S. Reed, M. Tzoufras, F.S. Tsung, L. Willingale, W.B. Mori, A. Rousse, K. Krushelnick, Z. Najmudin, Observation of synchrotron radiation from electrons accelerated in a petawatt-laser-generated plasma cavity. Phys. Rev. Lett. **100**, 105006 (2008)
41. S. Kneip, C. McGuffey, J. Martins, S. Martins, C. Bellei, V. Chvykov, F. Dollar, R. Fonseca, C. Huntington, G. Kalintchenko et al., Bright spatially coherent synchrotron x-rays from a table-top source. Nat. Phys. **6**, 980 (2010)
42. S. Kneip, C. McGuffey, J.L. Martins, M.S. Bloom, V. Chvykov, F. Dollar, R. Fonseca, S. Jolly, G. Kalintchenko, K. Krushelnick, A. Maksimchuk, S.P.D. Mangles, Z. Najmudin, C.A.J. Palmer, K.T. Phuoc, W. Schumaker, L.O. Silva, J. Vieira, V. Yanovsky, A.G.R. Thomas, Characterization of transverse beam emittance of electrons from a laser-plasma wakefield accelerator in the bubble regime using betatron x-ray radiation. Phys. Rev. ST Accel. Beams **15**, 021302 (2012)
43. M. Chen, A. Pukhov, T.P. Yu, Z.M. Sheng, Enhanced collimated gev monoenergetic ion acceleration from a shaped foil target irradiated by a circularly polarized laser pulse. Phys. Rev. Lett. **103**, 024801 (2009)
44. Z. Zhang, X. He, Z. Sheng, M. Yu, Hundreds mev monoenergetic proton bunch from interaction of 10^{20-21} W/cm^2 circularly polarized laser pulse with tailored complex target. Appl. Phys. Lett. **100**, 134103 (2012)
45. F. Zheng, H. Wang, X. Yan, T. Tajima, M. Yu, X. He, Sub-tev proton beam generation by ultra-intense laser irradiation of foil-and-gas target. Phys. Plasmas **19**, 023111 (2012)
46. M. Hegelich, S. Karsch, G. Pretzler, D. Habs, K. Witte, W. Guenther, M. Allen, A. Blazevic, J. Fuchs, J.C. Gauthier, M. Geissel, P. Audebert, T. Cowan, M. Roth, Mev ion jets from short-pulse-laser interaction with thin foils. Phys. Rev. Lett. **89**, 085002 (2002)
47. L. Willingale, S.P.D. Mangles, P.M. Nilson, R.J. Clarke, A.E. Dangor, M.C. Kaluza, S. Karsch, K.L. Lancaster, W.B. Mori, Z. Najmudin, J. Schreiber, A.G.R. Thomas, M.S. Wei, K. Krushelnick, Collimated multi-meV ion beams from high-intensity laser interactions with underdense plasma. Phys. Rev. Lett. **96**, 245002 (2006)
48. S.V. Bulanov, E.Y. Echkina, T.Z. Esirkepov, I.N. Inovenkov, M. Kando, F. Pegoraro, G. Korn, Unlimited ion acceleration by radiation pressure. Phys. Rev. Lett. **104**, 135003 (2010)

49. D.C. Carroll, O. Tresca, R. Prasad, L. Romagnani, P.S. Foster, P. Gallegos, S. Ter-Avetisyan, J.S. Green, M.J.V. Streeter, N. Dover, C.A.J. Palmer, C.M. Brenner, F.H. Cameron, K.E. Quinn, J. Schreiber, A.P.L. Robinson, T. Baeva, M.N. Quinn, X.H. Yuan, Z. Najmudin, M. Zepf, D. Neely, M. Borghesi, P. McKenna, Carbon ion acceleration from thin foil targets irradiated by ultrahigh-contrast, ultraintense laser pulses. New J. Phys. **12**, 045020 (2010)
50. S. Ter-Avetisyan, B. Ramakrishna, R. Prasad, M. Borghesi, P. Nickles, S. Steinke, M. Schnürer, K. Popov, L. Ramunno, N. Zmitrenko et al., Generation of a quasi-monoergetic proton beam from laser-irradiated sub-micron droplets. Phys. Plasmas **19**, 3112 (2012)
51. T. Toncian, M. Borghesi, J. Fuchs, E. d'Humières, P. Antici, P. Audebert, E. Brambrink, C. Cecchetti, A. Pipahl, L. Romagnani et al., Ultrafast laser-driven microlens to focus and energy-select mega-electron volt protons. Science **312**, 410 (2006)
52. I.A. Vartanyants, A. Singer, A.P. Mancuso, O.M. Yefanov, A. Sakdinawat, Y. Liu, E. Bang, G.J. Williams, G. Cadenazzi, B. Abbey, H. Sinn, D. Attwood, K.A. Nugent, E. Weckert, T. Wang, D. Zhu, B. Wu, C. Graves, A. Scherz, J.J. Turner, W.F. Schlotter, M. Messerschmidt, J. Lüning, Y. Acremann, P. Heimann, D.C. Mancini, V. Joshi, J. Krzywinski, R. Soufli, M. Fernandez-Perea, S. Hau-Riege, A.G. Peele, Y. Feng, O. Krupin, S. Moeller, W. Wurth, Coherence properties of individual femtosecond pulses of an x-ray free-electron laser. Phys. Rev. Lett. **107**, 144801 (2011)
53. C. Gutt, P. Wochner, B. Fischer, H. Conrad, M. Castro-Colin, S. Lee, F. Lehmkühler, I. Steinke, M. Sprung, W. Roseker, D. Zhu, H. Lemke, S. Bogle, P.H. Fuoss, G.B. Stephenson, M. Cammarata, D.M. Fritz, A. Robert, G. Grübel, Single shot spatial and temporal coherence properties of the slac linac coherent light source in the hard x-ray regime. Phys. Rev. Lett. **108**, 024801 (2012)
54. S. Cipiccia, M. Islam, B. Ersfeld, R. Shanks, E. Brunetti, G. Vieux, X. Yang, R. Issac, S. Wiggins, G. Welsh et al., Gamma-rays from harmonically resonant betatron oscillations in a plasma wake. Nat. Phys. **7**, 867 (2011)
55. H. Schwoerer, B. Liesfeld, H.-P. Schlenvoigt, K.-U. Amthor, R. Sauerbrey, Thomson-backscattered x rays from laser-accelerated electrons. Phys. Rev. Lett. **96**, 014802 (2006)
56. V. Karagodsky, D. Schieber, L. Schächter, Enhancing x-ray generation by electron-beam laser interaction in an optical bragg structure. Phys. Rev. Lett. **104**, 024801 (2010)
57. H. Schlenvoigt, K. Haupt, A. Debus, F. Budde, O. Jäckel, S. Pfotenhauer, H. Schwoerer, E. Rohwer, J. Gallacher, E. Brunetti et al., A compact synchrotron radiation source driven by a laser-plasma wakefield accelerator. Nat. Phys. **4**, 130 (2007)
58. M.-C. Chen, P. Arpin, T. Popmintchev, M. Gerrity, B. Zhang, M. Seaberg, D. Popmintchev, M.M. Murnane, H.C. Kapteyn, Bright, coherent, ultrafast soft x-ray harmonics spanning the water window from a tabletop light source. Phys. Rev. Lett. **105**, 173901 (2010)
59. S. Bagchi, P. Kiran, K. Yang, A. Rao, M. Bhuyan, M. Krishnamurthy, G. Kumar, Bright, low debris, ultrashort hard x-ray table top source using carbon nanotubes. Phys. Plasmas **18**, 014502 (2011)
60. Fair baseline technical report september 2006, http://www.gsi.de/en/research/fair.htm
61. LHC design report ch21, http://project-i-lhc.web.cern.ch/project-i-lhc/Reports.htm
62. F. Carminati, P. Foka, P. Giubellino, A. Morsch, G. Paic, J. Revol, K. Safarik, Y. Schutz, U. Wiedemann et al., Alice: physics performance report, volume i. J. Phys. G: Nucl. Part. Phys. **30**, 1517 (2004)

Chapter 4
Coherent Storage and Phase Modulation of Single Hard-X-Ray Photons

In this chapter, we change to another research direction of *manipulating single hard x-ray photon via magnetically controlling nuclear dynamics*. A pioneer experiment in this direction was performed by Y. Shvyd'ko et al. [1] in 1996. This work presents a scheme of storing a single nuclear excitation in a nuclear forward scattering setup using rotations of the hyperfine magnetic field. However, this setup does not achieve coherent storage, i.e., instead of photonic properties (phase, polarization, etc.), only the photon energy is transferred to the nuclear excitation. Because of this inconvenient feature, this pioneering research did not receive much attention from the quantum optics community [2] and remained at a fundamental level. Starting from the magnetic switching setup in Ref. [1], we develop here two designs that achieve coherent storage and π-phase modulation of single hard x-ray photons using nuclei. We theoretically proof the possibility of transferring not only the photon energy but also the photonic state to the nuclear excitation. This useful advantage may push this scheme to future applications, e.g., quantum information processes.

The chapter starts with the motivation of controlling an x-ray photon in Sect. 4.1.1 and an introduction to the so-called nuclear forward scattering in Sect. 4.1.2. In Sect. 4.1.3 we write down the used Maxwell-Bloch equations, and analyze the characteristic time scales of our system. Our main ideas are demonstrated in Sect. 4.2, starting with an intuitive picture used to explain the crucial physical phenomenon of quantum beats. This intuitive picture in fact triggers our main idea of how coherent storage can be achieved, and the numerical results are presented in the same section. In Sect. 4.3, we address the phase of a single hard x-ray photon with a two-target x-ray interferometer setup. We will learn a clear definition of the phase from the proposed measurements, and a way to continuously modulate the single-photon-phase.

W.-T. Liao, *Coherent Control of Nuclei and X-Rays*, Springer Theses,
DOI: 10.1007/978-3-319-02120-1_4,

4.1 Introduction and Motivation

4.1.1 Motivation of Controlling Hard X-Rays

Seeking for versatile solutions for quantum and classical computing on the most compact scale is one of the crucial objectives in both fundamental physics and information technology. The photon as a flying qubit is anticipated to be the fastest information carrier and provide the most efficient computing implementation. However, as showed in Fig. 4.1 extending Moore's law [3] to the future quantum photonic circuits must meet the bottleneck of the diffraction limit, i.e., few hundred nm for the optical region. Forwarding optics and quantum information to shorter wavelengths in the x-ray region has the potential of shrinking computing elements in future photonic devices such as the quantum photonic circuit [4]. For example, one can design a memory composed of a matrix of single atoms. As showed in Fig. 4.2a, if an optical laser beam is used as the access tool, the bit unit will include a large number of atoms. However, Fig. 4.2b demonstrates the potential of using a tightly focused hard x-ray to write/read data in a single atom, i.e., a single atom memory. Such a task requires mastery of x-ray optics and powerful control tools of single-photon wave packet amplitude, frequency, polarization and phase [5]. The development of x-ray optics elements has made already significant progress with the realization of x-ray diamond mirrors [6–8] and cavities [9], hard x-ray waveguides [10, 11] and the Fabry-Pérot resonator [12–14]. Efficient coherent photon storage for photon delay lines and x-ray phase modulation, preferably even for single-photon wave packets, are next milestones yet to be reached.

Moving towards the interactions in the x-ray regime [15–21], also new physical systems come into play, e.g., nuclei with low-lying collective states naturally arise as candidates for x-ray quantum optics studies. Nuclear quantum optics [22–25]

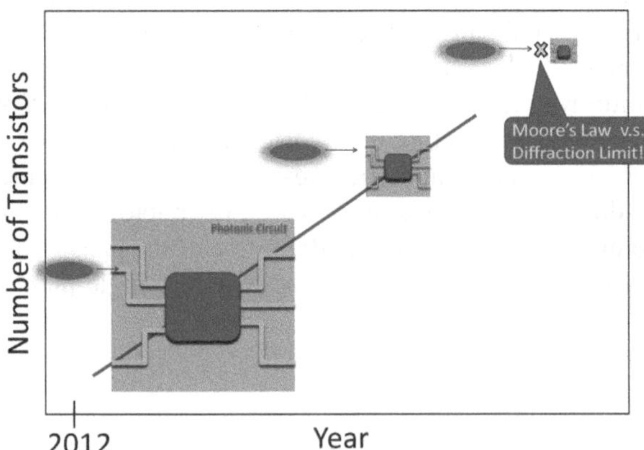

Fig. 4.1 The diffraction limit of optical photons will become the bottleneck of downsizing computing elements in future photonic devices

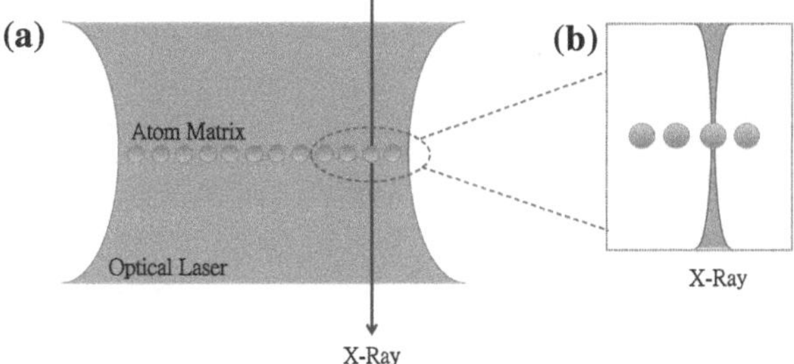

Fig. 4.2 A computing memory composed of a matrix of single atoms. **a** access the data by illuminating a number of atoms with a tightly focused optical laser. **b** the ability of treating a single atom as a bit unit using tightly focused hard x-ray beam to write/read data

and nuclear coherent population transfer [26] are rendered experimentally possible by the advent and commissioning of the x-ray free electron laser (XFEL) [27–29]. Coherent control tools based on nuclear cooperative effects [30–34] are known also from nuclear forward scattering experiments with third-generation light sources. The underlying physics here relies on the delocalized nature of the nuclear excitation produced by coherent XFEL or synchrotron radiation light, i.e., the formation of so-called nuclear excitons. Key examples in this direction is how manipulation of the hyperfine magnetic field in nuclear forward scattering systems provides means to store nuclear excitation energy [1] and in turn to generate keV single-photon entanglement [35].

4.1.2 Introduction to Nuclear Forward Scattering

The remarkable idea of nuclear forward scattering (NFS) with synchrotron radiation (SR) was firstly proposed by Stanley Ruby [32, 36, 37]. A typical NFS setup is showed in Fig. 4.3. Following S. Ruby's proposal, the conventional Mössbauer γ-ray sources were replaced by a SR beam. A nuclear target is placed in the SR beam line, and the forward-scattered photon signal is observed in the time domain instead of the energy spectra. The motivation for concentrating on the time spectra is crucial. In the original suggestion in 1974 [36], a key point was indicated "*...consider a scattering foil containing nuclei of mean life τ hit by the x-ray beam. The x-ray scattering, Compton scattering, photoelectric effects, etc. are prompt. But if some of the nuclei are excited by γ-ray absorption, the products (the scattered photon, or the conversion electron, or the X-ray following the internal conversion) of the subsequent nuclear decay will be delayed by about τ ...*". Therefore, provided the incident γ-ray pulse duration $\ll \tau$, one can discriminate the pure nuclear response from the background of electron-scattered photon simply by looking at different time windows with the setup in Fig. 4.3.

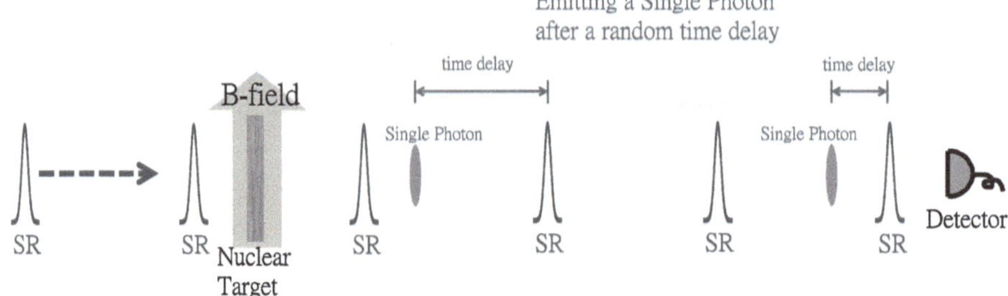

Fig. 4.3 Setup of nuclear forward scattering. A Mössbauer nuclear target is put in the beam line of synchrotron radiation under an external hyperfine magnetic field **B**, and a photon detector is located in the forward direction. A train of x-ray synchrotron radiation (SR) pulse continuously impinges on the nuclear target, and the detector registers the single photon signal in the time domain between two SR pulses

In measurements, the emitted photon number following the nuclear deexcitation is maximally one (more often none) due to the low photon degeneracy in the SR pulse and the narrow nuclear linewidth. This condition leads to a qualitative picture of shining a single resonant photon on an ensemble of resonant nuclei. One nucleus may be excited by the recoilless photon absorption, but there is no way to know which one from the measurements if neither nuclear spin flip nor internal conversion occurs. This delocalized single nuclear excitation is called *nuclear exciton* [32] or *nuclear polariton* [38], and was independently introduced by Hannon & Trammell and Kagan & Afanas'ev [32]. In atomic physics, the same concept is called single photon superradiance or directed spontaneous emission [39]. The collective wave-function of the nuclear exciton state is illustrated in Fig. 4.4 and can be written as [32, 39, 40]

$$|E\rangle = \frac{1}{\sqrt{N}} \sum_{\ell=1}^{N} e^{i\mathbf{k}_0 \cdot \mathbf{r}_\ell} |g\rangle |e_\ell\rangle, \tag{4.1}$$

where $|g\rangle|e_\ell\rangle$ denotes only the ℓth nucleus at position $\mathbf{r}_\ell$ is excited and all remaining $N - 1$ nuclei stay in the ground state. Furthermore, $\mathbf{k}_0$ is the wave vector of the incident resonant x-ray photon.

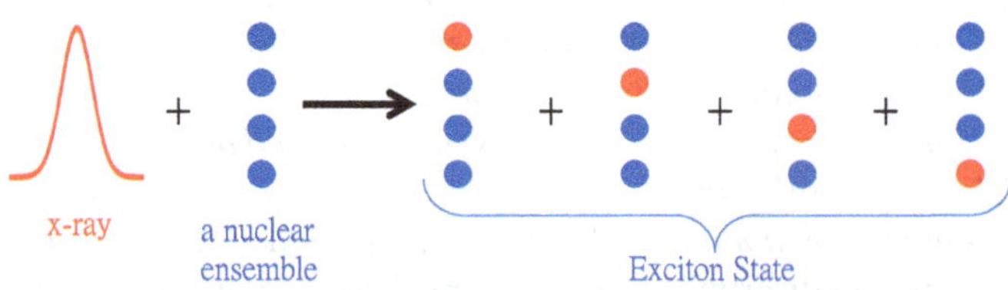

Fig. 4.4 A sketch of the nuclear exciton. The *red* Gaussian pulse is the incident x-ray, the *blue* (*red*) dots denote the nuclei in the ground (excited) state. The nuclear exciton is the superposition of all states for which one nucleus is excited and all remaining ones are in the ground state

The spontaneous decay rate of the nuclear exciton state $|E\rangle$ can be derived following the Weisskopf-Wigner theory [39, 41]. We assume that the nuclear ensemble is in the nuclear exciton state at $t = 0$. Thus, the complete nuclei-photon state is

$$|\psi\rangle = |G\rangle \otimes \sum_{\mathbf{k}} C_G^{\mathbf{k}}(t)|1\rangle_{\mathbf{k}} + C_E^{\mathbf{k_0}}(t)|E\rangle \otimes |0\rangle, \tag{4.2}$$

with the initial condition

$$C_G^{\mathbf{k}}(0) = 0,\ C_E^{\mathbf{k_0}}(0) = 1, \tag{4.3}$$

where $|G\rangle$ denotes all nuclei are in the ground state. At some later time the state $|\psi\rangle$ obeys the Schrödinger equation

$$\partial_t|\psi\rangle = -\frac{i}{\hbar}\widehat{H}_I|\psi\rangle \tag{4.4}$$

under the interaction Hamiltonian

$$\widehat{H}_I = \hbar\sum_{\ell,\mathbf{k}}\left[g_{\mathbf{k}}e^{i\mathbf{k}\cdot\mathbf{r}_\ell - i\Delta_k t}\widehat{\sigma}_+^\ell\widehat{a}_{\mathbf{k}} + g_{\mathbf{k}}e^{-i\mathbf{k}\cdot\mathbf{r}_\ell + i\Delta_k t}\widehat{\sigma}_-^\ell\widehat{a}_{\mathbf{k}}^\dagger\right]. \tag{4.5}$$

Here $\mathbf{k}$ and $\Delta_k = \omega_{\mathbf{k}} - \omega_N$ are the wave vector and the angular frequency detuning, respectively, of the interacting photon. $(\widehat{\sigma}_+^\ell, \widehat{\sigma}_-^\ell)$ are the nuclear (raising, lowering) operators for the nucleus at position r_ℓ, and $(\widehat{a}_{\mathbf{k}}^\dagger, \widehat{a}_{\mathbf{k}})$ are the photon (creation, annihilation) operators for a specific $\mathbf{k}$. $|1\rangle_{\mathbf{k}}$ and $|0\rangle$ denote the one photon Fock state of wave vector $\mathbf{k}$ and the vacuum state, respectively, and $g_{\mathbf{k}}$ is the nucleus-field coupling frequency.

From Eq. (4.4), we obtain the equation of motion for $C_G^{\mathbf{k}}$ and $C_E^{\mathbf{k_0}}$:

$$|G\rangle \otimes \sum_{\mathbf{k}}|1\rangle_{\mathbf{k}}\partial_t C_G^{\mathbf{k}} = -i\sum_{\ell,\mathbf{k}} g_{\mathbf{k}}e^{-i\mathbf{k}\cdot\mathbf{r}_\ell + i\Delta_k t}\widehat{\sigma}_-^\ell\widehat{a}_{\mathbf{k}}^\dagger C_E^{\mathbf{k_0}}|E\rangle \otimes |0\rangle, \tag{4.6}$$

$$|E\rangle \otimes |0\rangle\partial_t C_E^{\mathbf{k_0}} = -i\sum_{\ell,\mathbf{k}} g_{\mathbf{k}}e^{i\mathbf{k}\cdot\mathbf{r}_\ell - i\Delta_k t}\widehat{\sigma}_+^\ell\widehat{a}_{\mathbf{k}}|G\rangle \otimes \sum_{\mathbf{k}'} C_G^{\mathbf{k}'}|1\rangle_{\mathbf{k}'}. \tag{4.7}$$

After some simplifications, Eqs. (4.6) and (4.7) become

$$\begin{aligned}|G\rangle \otimes \sum_{\mathbf{k}}|1\rangle_{\mathbf{k}}\partial_t C_G^{\mathbf{k}} &= -\frac{i}{\sqrt{N}}|G\rangle \otimes \sum_{\ell,\mathbf{k}} g_{\mathbf{k}}e^{i(\mathbf{k}_0-\mathbf{k})\cdot\mathbf{r}_\ell + i\Delta_k t}|1\rangle_{\mathbf{k}}C_E^{\mathbf{k_0}}\\ &= -i\frac{\sqrt{N}}{V}(2\pi)^3|G\rangle \otimes \sum_{\mathbf{k}} g_{\mathbf{k}}e^{i\Delta_k t}|1\rangle_{\mathbf{k}}\delta^3(\mathbf{k}_0 - \mathbf{k})\,C_E^{\mathbf{k_0}}\end{aligned} \tag{4.8}$$

$$|E\rangle \otimes |0\rangle\partial_t C_E^{\mathbf{k_0}} = -i\sum_{\ell,\mathbf{k}} g_{\mathbf{k}}e^{i\mathbf{k}\cdot\mathbf{r}_\ell - i\Delta_k t}|g\rangle|e_\ell\rangle \otimes |0\rangle C_G^{\mathbf{k}}. \tag{4.9}$$

In Eq. (4.8), the atomic summation (for large number density as in a solid-state sample)

$$\sum_{\ell} e^{i(\mathbf{k}_0-\mathbf{k})\cdot\mathbf{r}_\ell} = \frac{N}{V}(2\pi)^3\,\delta^3(\mathbf{k}_0-\mathbf{k}) \tag{4.10}$$

is used [39] where V is the volume of the nuclear sample. Also, from Eq. (4.8), we obtain

$$C_G^{\mathbf{k}}(t) = -i\frac{\sqrt{N}}{V}(2\pi)^3\,g_{\mathbf{k}}\delta^3(\mathbf{k}_0-\mathbf{k})\int_0^t e^{i\Delta_k\tau}C_E^{\mathbf{k_0}}(\tau)d\tau. \tag{4.11}$$

On substituting Eq. (4.11) into (4.9), we obtain

$$|E\rangle\partial_t C_E^{\mathbf{k_0}} = -\frac{\sqrt{N}}{V}(2\pi)^3\sum_{\ell,\mathbf{k}}|g\rangle|e_\ell\rangle|g_{\mathbf{k}}|^2 e^{i\mathbf{k}\cdot\mathbf{r}_\ell-i\Delta_k t}\,\delta^3(\mathbf{k}_0-\mathbf{k})\int_0^t e^{i\Delta_k\tau}C_E^{\mathbf{k_0}}(\tau)d\tau. \tag{4.12}$$

Assuming (i) the wave number is nearly continuous for different field modes and (ii) the orientations of the nuclear transition dipole moment at different position $\mathbf{r}_\ell$ are the same, the summation can be replaced by an integral:

$$\sum_{\mathbf{k}} \to 2\frac{v}{(2\pi)^3}\int_{-\infty}^{\infty}dk_x\int_{-\infty}^{\infty}dk_y\int_{-\infty}^{\infty}dk_z \tag{4.13}$$

where v is the quantization volume, following from

$$|g_{\mathbf{k}}|^2 = \frac{\omega_k}{2\hbar\epsilon_0 v}P_{eg}^2\cos^2\theta, \tag{4.14}$$

where θ is the angle between the photon polarization and the transition dipole moment P_{eg}. Also, by using $\delta^3(\mathbf{k}_0-\mathbf{k}) = \delta(\mathbf{k}_{0,x}-\mathbf{k}_x)\,\delta(\mathbf{k}_{0,y}-\mathbf{k}_y)\,\delta(\mathbf{k}_{0,z}-\mathbf{k}_z)$, Eq. (4.12) becomes

$$\begin{aligned}
|E\rangle\partial_t C_E^{\mathbf{k_0}} &= -\frac{\sqrt{N}}{\hbar\epsilon_0 V}P_{eg}^2\cos^2\theta\sum_{\ell}|g\rangle|e_\ell\rangle\int_{-\infty}^{\infty}dk_x\int_{-\infty}^{\infty}dk_y\int_{-\infty}^{\infty}dk_z e^{i\mathbf{k}\cdot\mathbf{r}_\ell}\\
&\quad\times\omega_k\delta(\mathbf{k}_{0,x}-\mathbf{k}_x)\,\delta(\mathbf{k}_{0,y}-\mathbf{k}_y)\,\delta(\mathbf{k}_{0,z}-\mathbf{k}_z)\int_0^t e^{-i\Delta_k(t-\tau)}C_E^{\mathbf{k_0}}(\tau)d\tau\\
&= -\frac{\sqrt{N}\omega_{k_0}}{\hbar\epsilon_0 V}P_{eg}^2\cos^2\theta\sum_{\ell}|g\rangle|e_\ell\rangle e^{i\mathbf{k_0}\cdot\mathbf{r}_\ell}\int_0^t e^{-i\Delta_{k_0}(t-\tau)}C_E^{\mathbf{k_0}}(\tau)d\tau\\
&= -\frac{N\omega_{k_0}}{\hbar\epsilon_0 V}P_{eg}^2\cos^2\theta|E\rangle\int_0^t e^{-i\Delta_{k_0}(t-\tau)}C_E^{\mathbf{k_0}}(\tau)d\tau
\end{aligned} \tag{4.15}$$

Finally, the equation of motion for $C_E^{\mathbf{k_0}}$ is obtained:

$$\partial_t^2 C_E^{\mathbf{k_0}} = -\left(\frac{N\omega_{k_0}}{\hbar\epsilon_0 V} P_{eg}^2 \cos^2\theta\right) C_E^{\mathbf{k_0}}, \tag{4.16}$$

and the solution of Eq. (4.4) is [42, 43]

$$\begin{aligned} C_E^{\mathbf{k_0}}(t) &= \cos\left(t P_{eg} \cos\theta \sqrt{\frac{N\omega_{k_0}}{\hbar\epsilon_0 V}}\right) \\ &= \cos\left(\sqrt{N}\Omega_{k_0} t\right) \end{aligned} \tag{4.17}$$

$$C_G^{\mathbf{k_0}}(t) = \sin\left(\sqrt{N}\Omega_{k_0} t\right) \tag{4.18}$$

$$C_G^{\mathbf{k}\neq\mathbf{k_0}}(t) = 0. \tag{4.19}$$

From Eq. (4.18), the probability of finding the nuclear exciton is

$$|C_E^{\mathbf{k_0}}(t)|^2 = \cos^2\left(\sqrt{N}\Omega_{k_0} t\right), \tag{4.20}$$

and that of finding one photon in the system is

$$|C_G^{\mathbf{k_0}}(t)|^2 = \sin^2\left(\sqrt{N}\Omega_{k_0} t\right). \tag{4.21}$$

Here, we summarize the properties of the nuclear exciton state:

- The forward emission: the nuclear exciton is excited by one photon absorption with the wave vector $\mathbf{k_0}$, and emits a photon later only in the same direction of $\mathbf{k_0}$ (see Eqs. (4.18) and (4.19)) if the nuclear number N is large enough so that Eq. (4.10) is applicable.
- Collective oscillation: in free space the spontaneous decay behavior of a single excited nucleus can be derived from the Weisskopf-Wigner theory. Remarkably, with the same theory a large number of nuclei in the nuclear exciton state results in the collective oscillation described by Eq. (4.17). The oscillation can be understood as the emitted photon being re-absorbed and re-emitted in the nuclear ensemble [43], i.e., multiple scattering.

However, Eq. (4.20) does not initially decay with the same rate as that measured in experiments [32]. Thus, the concept of the nuclear exciton or superradiance only qualitatively explains the NFS time spectra. To get the full description of NFS one must refer to the wave property of the radiation and consider the so-called coherent light pulse propagation through the nuclear resonant medium (see Sect. 2.2.1) [30, 32]. The following sections are devoted to an investigation of our system by using this semi-classical approach.

4.1.3 Model and Time Scales

Our considered NFS system is illustrated in Fig. 4.5. In Fig. 4.5a, a ^{57}Fe enriched solid-state target (the green cuboid) under the action of a static magnetic field (the red arrow) is bombarded by a train of hard x-ray pulses (the blue arrow). The x-ray propagates in the y-direction and its polarization is in the x-direction. The magnetic field is pointed to the z-direction set as the quantization axis. Figure 4.5b shows the hyperfine levels under the Zeeman splitting. Two $\Delta m = m_e - m_g = 0$ transitions are driven by the x-ray pulse. The dynamics of the density matrix $\widehat{\rho}(t)$ is described by the Maxwell-Bloch equations [25, 44–46]:

$$\begin{aligned} \partial_t \widehat{\rho} &= \frac{1}{i\hbar}\left[\widehat{H}, \widehat{\rho}\right] + \widehat{\rho}_s \,, \\ \frac{1}{c}\partial_t \Omega + \partial_y \Omega &= i\eta \left(a_{31}\rho_{31} + a_{42}\rho_{42}\right), \end{aligned} \tag{4.22}$$

with the interaction Hamiltonian

$$\widehat{H} = -\frac{\hbar}{2}\begin{pmatrix} 2\Delta_g & 0 & a_{31}\Omega^* & 0 \\ 0 & -2\Delta_g & 0 & a_{42}\Omega^* \\ a_{31}\Omega & 0 & -2(\Delta + \Delta_e) & 0 \\ 0 & a_{42}\Omega & 0 & -2(\Delta - \Delta_e) \end{pmatrix}.$$

The initial and boundary conditions are

$$\rho_{ij}(0, y) = \frac{1}{2}\left(\delta_{i1}\delta_{1j} + \delta_{i2}\delta_{2j}\right) ,$$

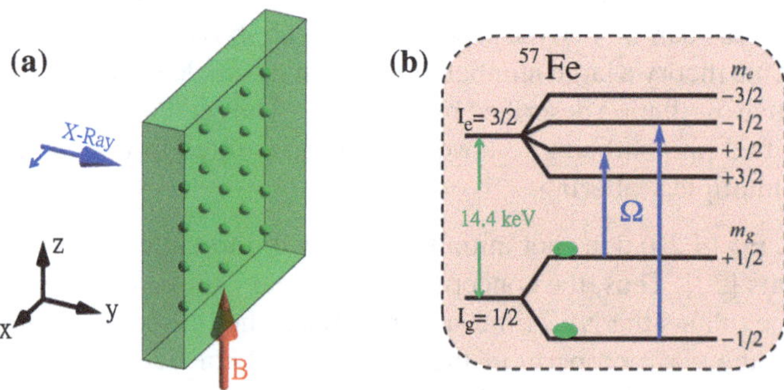

Fig. 4.5 **a** Sketch of the considered NFS system. The *red arrow* denotes an external magnetic field **B**, the *blue arrow* is the linearly polarized x-ray and the *green cuboid* matrix depicts the nuclear target. **b** The Zeeman sub-level scheme of ^{57}Fe nucleus under an external magnetic field **B**. Ω is the Rabi frequency of an incident x-ray (*blue arrows*), $I_{g(e)}$ is the spin of nuclear ground (excited) state, and $m_{g(e)}$ denotes the projection of nuclear ground (excited) state spin on the z-axis

$$\Omega(0, y) = 0,$$
$$\Omega(t, 0) = \varphi Exp\left[-\left(\frac{t}{T}\right)^2\right]. \tag{4.23}$$

In the equations above Δ is the x-ray detuning to the 14.4 keV transition assumed to be zero and $\Delta_{g(e)}$ denotes the Zeeman energy splitting of the nuclear ground (excited) state proportional to the magnetic field **B**. In Eq. (4.22), $\rho_{ij} = A_i A_j^*$ are the density matrix elements of $\widehat{\rho}$ for the nuclear wave function $|\psi\rangle = A_1|\frac{1}{2}, -\frac{1}{2}\rangle + A_2|\frac{1}{2}, \frac{1}{2}\rangle + A_3|\frac{3}{2}, -\frac{1}{2}\rangle + A_4|\frac{3}{2}, \frac{1}{2}\rangle$. The ket vectors $|I, m\rangle$ are the eigenvectors of the two ground and two excited states hyperfine levels[1] where I is the spin of nuclear state, and m denotes the projection of nuclear spin on the quantization axis. Furthermore, $a_{31} = a_{42} = \sqrt{2/3}$ are the corresponding Clebsch-Gordan coefficients [46, 47] for the $\Delta m = 0$ transitions and $\widehat{\rho_s}$ describes the spontaneous decay [39]. The parameter η is defined as $\eta = \frac{6\Gamma}{L}\xi$, where $\Gamma = 1/141.1\,\text{GHz}$ is the spontaneous decay rate of excited states, ξ represents the effective resonant thickness [44, 45, 47] and $L = 10\,\mu\text{m}$ the thickness of the target, respectively. Further notations are Ω for the Rabi frequency which is proportional to the electric field **E** of the x-ray pulse [39, 46], the incident probe chosen such that $\varphi \ll \Gamma$ to prevent any Rabi oscillation[2] with the pulse duration $T = 0.5\,\text{ns}$. δ_{ij} is the Kronecker delta and c the speed of light.

Figure 4.6 illustrates a typical unperturbed time spectrum of NFS. We numerically solve Eq. (4.22) with the corresponding parameters $\xi = 10$ and the Zeeman shift $\Delta_B = \Delta_e + \Delta_g = 30\Gamma$. Our density matrix calculations have been double-checked via the comparison with results from the iterative solution of the wave equations originally proposed by Shvyd'ko [1]. The agreement is complete for both electric field envelope and scattered light intensity, proving the equivalence of the two calculation methods. Blue solid lines are the intensities of the NFS signal proportional to $|\Omega(t, L)|^2$ and red dashed lines denote qualitatively the applied hyperfine magnetic field. For comparison, we present also the spontaneous decay curve (gray dotted line) proportional to $e^{-\Gamma t}$. Two special features of NFS time spectra are the dynamical beat and the quantum beat [47, 48]. The dynamical beat results from the dispersion relation together with the broadband excitation [30, 44] for a single nuclear transition, and it also can be explained by the multiple scattering [31, 45]. The analytic solution of dynamical beat is

$$\left(\frac{\xi}{\sqrt{\xi\Gamma t}} J_1\left[2\sqrt{\xi\Gamma t}\right]\right)^2 e^{-\Gamma t}, \tag{4.24}$$

[1] We neglect the two excited states $|\frac{3}{2}, -\frac{3}{2}\rangle$ and $|\frac{3}{2}, \frac{3}{2}\rangle$ which are not involved in the whole scattering process due to the polarization of the incident x-ray.

[2] The choice of φ and T is not so strict. In the whole chapter we calculate the normalized time spectra $\Omega(t, z)/\varphi$ which are φ independent under the condition of $\varphi \ll \Gamma$. The pulse duration T is chosen such that $T \ll 1/\Gamma$ to separate the pure nuclear response from the transient effects (equally, $1/T \gg \Gamma$ to fit the situation of broadband excitation).

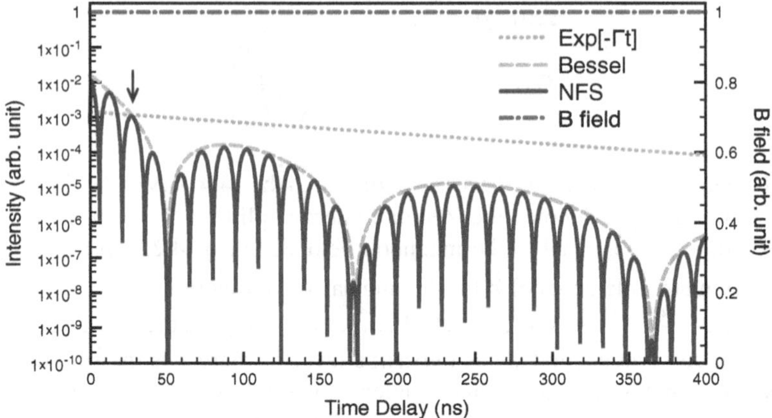

Fig. 4.6 The unperturbed NFS time spectrum. The *gray dotted line* is the spontaneous decay, while the *red dashed-dotted line* depicts the static magnetic field **B**. The *blue solid line* depicts the NFS time spectrum under the action of the static magnetic field **B**, and the frequent beating pattern is labeled as the quantum beat. The *green dashed line* shows the dynamical beat described by Eq. (4.24), and it is the NFS time spectrum without Zeeman splitting

where J_1 is the Bessel function of first kind [30, 31, 44, 45, 47] (see Sect. 2.2 for the derivation). When the Zeeman splitting is induced by the hyperfine magnetic field, the quantum beat is superimposed to the dynamical beat, as showed by the blue solid line. In Fig. 4.6, we identify three characteristic time scales in a typical NFS system:

1. The time scale of coherent effects is limited by decoherence process, e.g., the spontaneous decay in our case. So we can expect that the coherent signal will be observed within the lifetime of nuclear excited state ($1/\Gamma = 141.1$ ns).
2. The period of the dynamical beat corresponds to the superradiant speed-up effect resulting in a time scale of $1/(\xi\Gamma)$.
3. The quantum beat is caused by the interference between emitted photons from two deexcitations of $|\frac{3}{2}, -\frac{1}{2}\rangle \rightarrow |\frac{1}{2}, -\frac{1}{2}\rangle$ and $|\frac{3}{2}, \frac{1}{2}\rangle \rightarrow |\frac{1}{2}, \frac{1}{2}\rangle$, therefore the Zeeman shift $\Delta_B = \Delta_e + \Delta_g$ gives a time scale of of $1/\Delta_B$.

In the whole chapter, we will consider the condition of $1/\Gamma > 1/\xi\Gamma > 1/\Delta_B$ to prevent hybrid beats [47] and show that one can manipulate the long-time-scale phenomena in $1/(\xi\Gamma)$ by controlling the short-time-scale effect in $1/\Delta_B$.

One may argue that the nuclear exciton state is missing in the semi-classical Maxwell-Bloch equations, and the collective effect is put in by introducing the factor η in Eq. (4.22). This semi-classical approach turns out to work very well [1, 30, 31, 44, 45, 47, 48]! Three aspects are important here:

- The exciton state [32] (time-Dicke state [39]) in Eq. (4.2) only explains the forward emission, whereas it is not able to give the correct time spectra, e.g., the complete dynamical beat pattern.

- The only ansatz of the Maxwell-Bloch equations we use is the forward emission which is phenomenological. So far, this ansatz seems to be correct in NFS and gives the complete features of time-delay single photon statistics.
- Why does the classical field correctly describe the behavior of single photons? This is an important question. This would be the case if the photon state that we are discussing within the time-delay window was a *coherent state*, i.e., $|P\rangle = C_0|0\rangle + C_1|1\rangle + C_2|2\rangle + \ldots$, where $|n\rangle$ is the n photon Fock state and $|C_0|^2 \gg |C_1|^2 \gg |C_2|^2 \gg \ldots$. After the incident pulse passes through a nuclear target, mostly no nuclear response is detected, and only few single-photon events are registered, such that the coherent state condition is fulfilled. For that reason we can describe our experiments with classical formulas [49].

4.2 Coherent Storage and π Phase Modulation

In this section, we present two schemes to manipulate single hard x-ray photons using resonant scattering of light off nuclei in a nuclear forward scattering setup. Using sychrotron radiation, the formation of a nuclear exciton consisting of a single delocalized excitation opens the possibility to control the coherent decay and therefore emission of the scattered photon. Making use of this feature, we first put forward how to coherently store a single hard x-ray photon for time intervals of 10–100 ns by turning off the hyperfine magnetic field in a NFS system. The stored single photon can be released by turning on the magnetic field. We emphasize that our scheme conserves not only the excitation energy, as already pioneeringly demonstrated in Ref. [1], but also the photonic polarization and phase beyond the ps time range. Next, we demonstrate how to modulate the single photon with a phase shift of π by reversing the hyperfine magnetic field orientation at proper time points.

4.2.1 Results and Discussion

Since the quantum beat plays a very important role in our scheme of manipulating a single x-ray photon, we explain its origin in detail with Fig. 4.7. Figure 4.7 schematically illustrates the time evolution of our NFS system. The external magnetic field **B**, depicted by the red line, is present before the x-ray pulse impinges on the target at T_0. The orange rotating arrows depict the time evolution of the nuclear transition current matrix elements as defined in Ref. [1]. In our treatment, this is equivalent with investigating the coherence terms $i\rho_{42}$ and $i\rho_{31}$ [44, 45]. Initially, the ensemble of ^{57}Fe nuclei is excited by the x-ray pulse at T_0. Subsequently, the purely real[3] currents

[3] The initial phases of the nuclear currents are arbitrary, and we assume a zero phase for convenience. In the next section we will see that the absolute phase cannot be defined, and only the relative phase is meaningful.

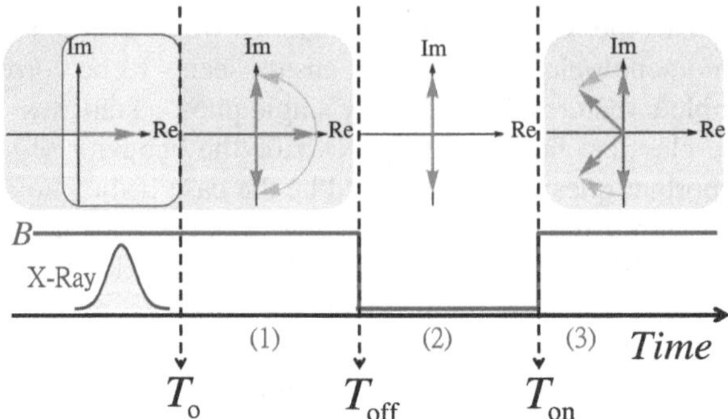

Fig. 4.7 The mechanism of NFS quantum beat and the storage scheme for a hard x-ray photon. The *rotating orange arrows* are the nuclear currents, the *red line* depicts the external magnetic field **B** and the *blue* Gaussian pulse illustrates the incident x-ray

are abruptly built. In the time interval (1), the two currents start to rotate in opposite directions on the complex plane with the factor of $e^{\pm i \Delta_B t}$ caused by the Zeeman shifts. The corresponding phase gain is $\pm\Delta_B t$, where t is the time duration after the nuclei were excited. During the whole excited nuclear evolution, each maximum of NFS time spectra appears when two nuclear currents constructively superpose on the real axis. On the other hand, each minimum happens when two currents destructively cancel each other on the imaginary axis. This simple picture of the quantum beat gives us the intuitive idea of storing single hard x-ray photon via turning off the magnetic field at a any minimum of the NFS time spectra, i.e., to freeze the nuclear currents on the imaginary axis.[4] We briefly explain our photon storage scheme in the rest of Fig. 4.7. At T_{off} the **B** field is turned off and later turned back on at T_{on}. Within the time interval (2), the quantum beat arising from the interference between the two $\Delta m = 0$ transitions is frozen with the factor of $e^{\pm i \Delta_B \tau}$ since the hyperfine field has vanished. During the time interval (3), the presence of the magnetic field makes the quantum beat emerge again.

In Fig. 4.8 we show the numerical solution of Eq. (4.22) with the same parameters of Fig. 4.6 to demonstrate our photon storage scheme by turning off the magnetic field at $t = 21$ ns. Both nuclear currents corresponding to the $\Delta m = 0$ transitions are frozen on the imaginary axis and present destructive interference. In this case the intensity of the emitted radiation is significantly suppressed, being three orders of magnitude smaller compared to the unperturbed spectrum in Fig. 4.6. Later on by turning the hyperfine magnetic field on again at (a) $t = 75$ ns; (b) $t = 125$ ns; (c) $t = 175$ ns, the photon signal is observed again within the time interval (3) of Fig. 4.7. Figure 4.8 also shows that the stored nuclear excitation energy experiences spontaneous decay during the storage time [1]. We indicate this effect by monitoring

[4] This means that no *coherent photon* is emitted due to the destructive quantum interference. However, incoherent photons due to the spontaneous decay may still be emitted, limiting the time scale of coherent effects as mentioned before.

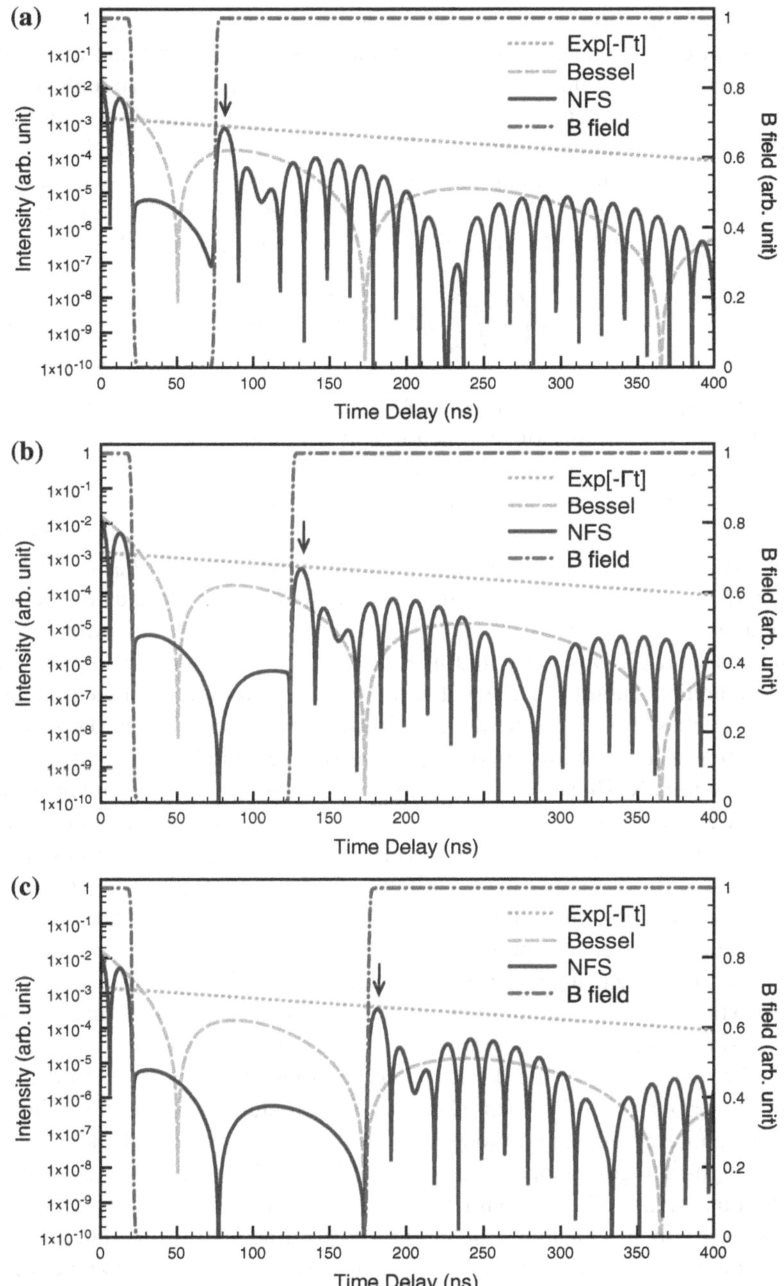

Fig. 4.8 The storage of hard x-ray single photons. The magnetic field **B** is turned off at 21 ns and then turned on at **a** 75 ns, **b** 125 ns and **c** 175 ns. All coloured lines share the same legend as Fig. 4.6. The *purple arrows* point out the first stored peaks after the nuclear signal retrieval

the stored first signal peak, pointed out by a purple arrow, which follows the gray dotted line of $e^{-\Gamma t}$ in each figure. One should however keep in mind that with SR we do not store single photons in a deterministic fashion. In experiments based for example on the situation presented in Fig. 4.8b, one will mostly observe the photon counts during two time windows $t < 21$ ns and $t > 125$ ns. Each SR pulse will mostly trigger no nuclear signal, and the nuclear scattering events will occur in a statistic way with a relatively low probability. However, this may be improved by using the fully coherent XFEL to fulfill the condition mentioned in Sect. 3.1:

$$\sin\left(\frac{1}{2}\int_{-\infty}^{\infty}\Omega^{\text{eff}}(t)dt\right) = \frac{1}{\sqrt{N}}. \tag{4.25}$$

One can adjust either the effective Rabi frequency $\Omega^{\text{eff}}(t)$ of a fully coherent XFEL pulse or the number of the nuclei N to excite the nuclear exciton state per XFEL shot in a nearly deterministic fashion.

The most significant advantage of our storage scheme is the conservation of the photonic polarization and phase. The electric field envelopes of the scattered photon are presented in Fig. 4.10. In Fig. 4.10a, the magnetic field $\mathbf{B}$ before $T_{\text{off}} = 80.5$ ns and that after $T_{\text{on}} = 175$ ns are the same and the phase before storage and after retrieving is continuous. Storage of nuclear excitation energy by magnetic field rotations in NFS experiments with SR was presented in Ref. [1]. However, since in this pioneering work the magnetic Hamiltonian is not zero during the storage, both the polarization [50] and the phase of the particular polarization components cannot be stored and the properties of the released photon depend on the switching instants. With the advent of coherent XFEL sources and x-ray quantum optics and quantum information experiments, phase storage and modulation become crucial for many applications. So far, coherent trapping of hard x-rays in crystal cavities provides photon storage for time intervals in the ps range [9]. Our scheme provides robust phase and polarization storage of the x-ray photon on the 10–100 ns scale determined by the nuclear lifetime.

In order to implement our photon storage scheme experimentally, a material with no intrinsic nuclear Zeeman splitting like stainless steel $Fe_{55}Cr_{25}Ni_{20}$ [51, 52] is required. The remaining challenge is to turn off and on the external magnetic fields of few Tesla on the ns time scale. According to our calculations for the case of Fig. 4.8, the raising time of the $\mathbf{B}$ field should be shorter than 50 ns (the raising time was considered 4 ns for all presented cases). This could be achieved by using small single- or few-turn coils and a moderate pulse current of approx. 15 kA from low-inductive high-voltage "snapper" capacitors [53]. Another mechanical solution, e.g., the lighthouse setup [32, 54] could be used to bring the excited target out and in a region with confined static magnetic field $\mathbf{B}$. With the setup illustrated in Fig. 4.9, we estimate that a rotor with rotational frequencies ω_R of up to 70 kHz and a diameter of few mm [32] is fast enough to rotate the sample out a depth of few μm in a few tens of ns.

Let us turn to the phase modulation of an single x-ray photon. A case of π phase modulation is presented in Fig. 4.10b by using the same parameters those used in

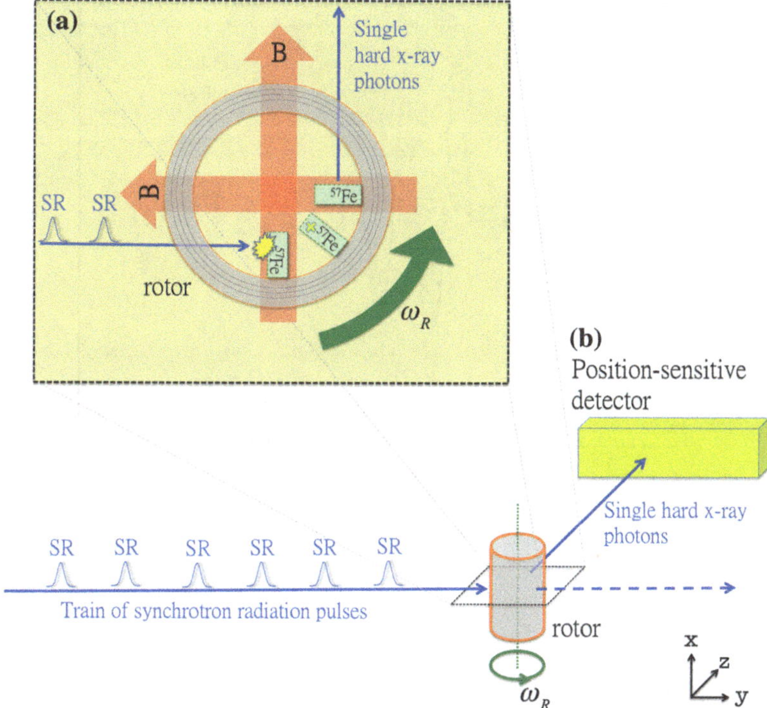

Fig. 4.9 Sketch of the lighthouse setup for the coherent storage of hard x-ray single photons. **a** Bird view of the lighthouse setup. *Gray area* depicts the side view of the rotor rotating with angular frequency ω_R, the two *red wide arrows* illustrate the regions with confined static magnetic field **B** and the *blue arrows* the trajectories of SR and emitted single hard x-ray photons. The *light green* rectangles depict snapshots of the rotating ^{57}Fe target attached on the inner surface of the rotor. **b** The geometric arrangement of the lighthouse scheme

Fig. 4.10a. If, however, the retrieving magnetic field is applied in opposite direction as shown in Fig. 4.10b, the phase of the released photonic wave packet will be modulated with a shift of π. This is caused by the effect of reversed time related with the change of sign of the hyperfine magnetic field [55, 56], i.e., all the nuclear currents evolve backwards in time. However, since the phase of a single photon is totally undefined, a method for measuring the magnetically modulated π phase shift is to be addressed. For the definition and the measurement of this π-phase shift of single photons, we refer to the echo technique using two nuclear targets [38, 51, 52] and demonstrate for the first time a magnetically induced nuclear exciton echo without any mechanical vibration of the targets. In the next section, we will see how this feasible echo two-target setup can also be used for photon storage involving a mere rotation of the hyperfine magnetic field by 180°.

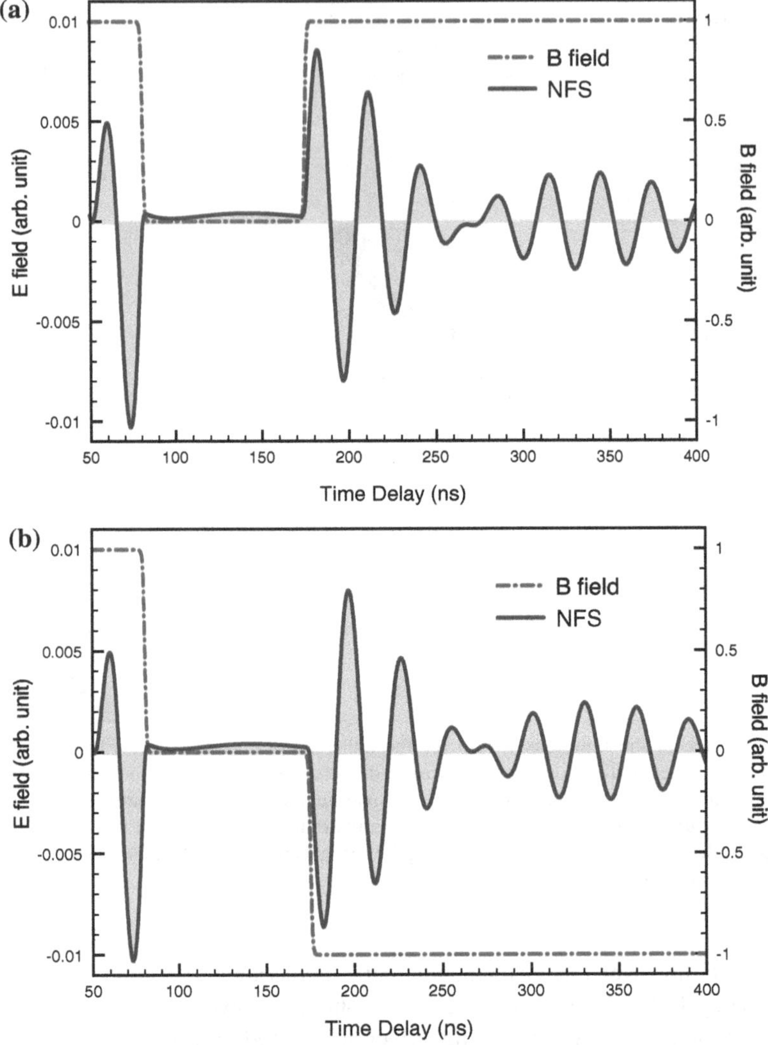

Fig. 4.10 The π phase modulation of a single hard x-ray photon under the action of the magnetic field **B**. The two figures show the electric field of the coherently scattered x-ray. The magnetic field is turned off at 80.5 ns and then turned on at 175 ns in the (**a**) same and (**b**) opposite direction

4.3 Hard X-Ray Interferometer and the Phase of a Single Photon Wavepacket

To unambiguously define the phase shift of a single photon, one must refer to an interferometer. A typical x-ray optics setup would require to let the π-modulated photon interfere with a part of the original pulse on a triple Laue interferometer

[57, 58]. We adopt here another approach, namely, the simple and elegant photon echo solution used in NFS experiments with SR [38, 51, 52, 59, 60] to allow the scattered photon to interfere with itself in the two-target setup presented in Fig. 4.11. A reversible magnetic field $\mathbf{B}_1$ is applied to target 1, and an irreversible $\mathbf{B}_2$ is applied to target 2. The target response is determined by

$$R(\alpha, \Delta_B, t) = \delta(t) - W(\xi, \Delta_B, t), \tag{4.26}$$

and [32, 38, 60]

$$W(\alpha, \Delta_B, t) = \frac{\xi}{\sqrt{\xi \Gamma t}} J_1\left(2\sqrt{\xi \Gamma t}\right) e^{-\frac{\Gamma}{2}t + i\Delta_B t}, \tag{4.27}$$

and the forward-scattered x-ray field is then given by [38]

$$E^{(1)}(t) = \int_0^t R(\xi, \Delta_B, t-\tau) E^{(0)}(\tau) d\tau. \tag{4.28}$$

Using $E^{(0)}(t) = \delta(t)$ as the incident x-ray, the resulting electric field is the real part of

$$\begin{aligned} E^{(2)}(t) = \delta(t) &- W(\xi_1, \Delta_{B1}, t) - W(\xi_2, \Delta_{B2}, t) \\ &+ \int_0^t W(\xi_2, \Delta_{B2}, t-\tau) W(\xi_1, \Delta_{B1}, \tau) d\tau. \end{aligned} \tag{4.29}$$

This depicts the interference of four possible coherent scattering channels [32, 38]:

1. $\delta(t)$, the incident pulse produces no nuclear scattering.
2. $-W(\xi_1, \Delta_{B1}, t)$, the photon is scattered by target 1 only.
3. $-W(\xi_2, \Delta_{B2}, t)$, the photon is scattered by target 2 only.

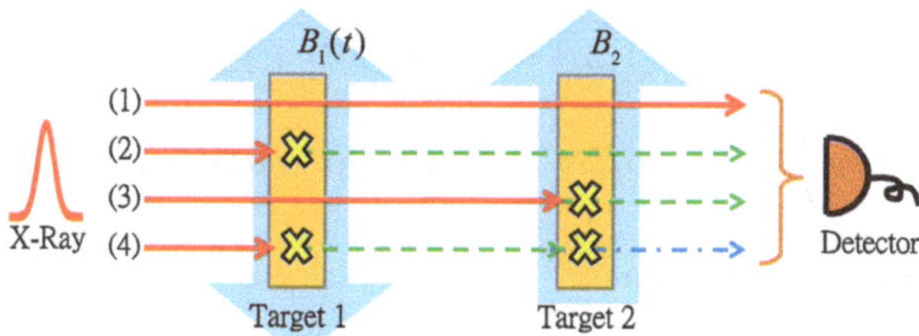

Fig. 4.11 A two-target x-ray interferometer setup. Four coherent scattering channels of the x-ray photons can interfere. *Yellow crosses* illustrate the formation of the nuclear exciton. The *light blue* vertical wide arrows show the applied magnetic fields: a reversible $\mathbf{B}_1$ is applied to target 1, whereas an irreversible $\mathbf{B}_2$ is applied to target 2

4. the mutual integral, i.e., the photon is first scattered by target 1 and subsequently by target 2.

Channel (2) and (3) cancel each other out when the effective thicknesses of the two targets are equal $\xi_1 = \xi_2$ and $\mathbf{B}_1(t > T_{\text{on}}) = -\mathbf{B}_2$, i.e., $\mathbf{B}_1(t)$ is reversed at $t = T_{\text{on}}$. Hence a significant suppression of the NFS signal can serve as signature for the effective π phase shift magnetically modulated in target 1. The statement "phase of single hard x-ray photon" is not strictly correct. A more appropriate one is "phase of single hard x-ray photon wavepacket", and the wavepacket meant here is $E^{(2)}(t)$ in the considered cases. The phase that can be modulated is the relative value between different scattering channels. In the setup presented in Fig. 4.11, without any magnetic switching one cannot distinguish channels (2) and (3) since they are identical. However, once the orientation of one magnetic field, e.g., $\mathbf{B}_1$ is reversed at a node time, destructive interference happens, and this operation provides the unambiguous definition of our phase. On the other hand, the amplitude of each channel is described by each component of the wavepacket $E^{(2)}(t)$, and therefore one can term it as "phase of single hard x-ray photon wavepacket".

Inspired by Eq. (4.29), we describe the two-target setup by the following Maxwell-Bloch equations [25]:

$$\partial_t \widehat{\alpha} = \frac{1}{i\hbar}\left[\widehat{H}_1, \widehat{\alpha}\right] + \widehat{\alpha}_s\,,$$
$$\frac{1}{c}\partial_t \Omega_1 + \partial_y \Omega_1 = i\eta\left(a_{31}\alpha_{31} + a_{42}\alpha_{42}\right), \tag{4.30}$$

with the interaction Hamiltonian

$$\widehat{H}_1 = -\frac{\hbar}{2}\begin{pmatrix} 2\Delta_{g1} & 0 & a_{31}\Omega_1^* & 0 \\ 0 & -2\Delta_{g1} & 0 & a_{42}\Omega_1^* \\ a_{31}\Omega_1 & 0 & -2(\Delta+\Delta_{e1}) & 0 \\ 0 & a_{42}\Omega_1 & 0 & -2(\Delta-\Delta_{e1}) \end{pmatrix},$$

and

$$\partial_t \widehat{\beta} = \frac{1}{i\hbar}\left[\widehat{H}_2, \widehat{\beta}\right] + \widehat{\beta}_s\,,$$
$$\frac{1}{c}\partial_t \Omega_2 + \partial_y \Omega_2 = i\eta\left(a_{31}\beta_{31} + a_{42}\beta_{42}\right), \tag{4.31}$$

with the interaction Hamiltonian

$$\widehat{H}_2 = -\frac{\hbar}{2}\begin{pmatrix} 2\Delta_{g2} & 0 & a_{31}\Omega_2^* & 0 \\ 0 & -2\Delta_{g2} & 0 & a_{42}\Omega_2^* \\ a_{31}\Omega_2 & 0 & -2(\Delta+\Delta_{e2}) & 0 \\ 0 & a_{42}\Omega_2 & 0 & -2(\Delta-\Delta_{e2}) \end{pmatrix}.$$

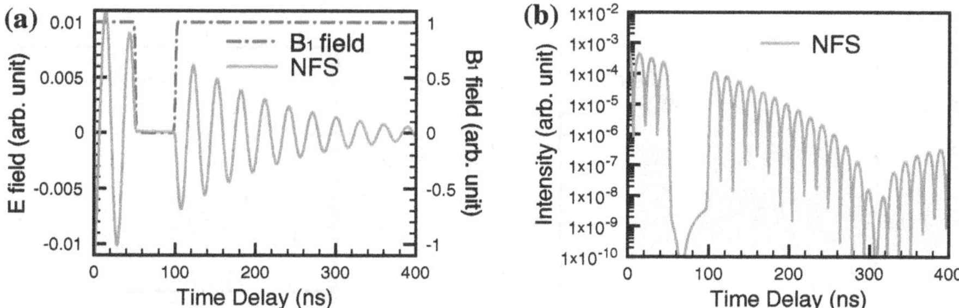

Fig. 4.12 **a** The electric field of the x-ray scattered from target 1 only. **b** The resulting NFS signal registered by the detector in Fig. 4.11. The magnetic field $\mathbf{B}_1$ and $\mathbf{B}_2$ are turned off at 51 ns and on at 100 ns in the original direction

The initial and boundary conditions are

$$
\begin{aligned}
\alpha_{ij}(0, y) &= \frac{1}{2}\left(\delta_{i1}\delta_{1j} + \delta_{i2}\delta_{2j}\right), \\
\Omega_1(0, y) &= 0, \\
\Omega_1(t, 0) &= \varphi Exp\left[-\left(\frac{t}{T}\right)^2\right], \\
\beta_{ij}(0, y) &= \frac{1}{2}\left(\delta_{i1}\delta_{1j} + \delta_{i2}\delta_{2j}\right), \\
\Omega_2(0, y) &= 0, \\
\Omega_2(t, 0) &= \Omega_1(t, L).
\end{aligned}
\tag{4.32}
$$

In the equations above, $\widehat{\alpha}(t)$ and $\widehat{\beta}(t)$ are now the nuclear density matrices for target 1 and target 2, respectively. All other notations are equivalent to those in Eq. (4.29). The Zeeman shift notations $\Delta_{B1} = \Delta_{g1} + \Delta_{e1}$ and $\Delta_{B2} = \Delta_{g2} + \Delta_{e2}$ are used for convenience.

We numerically solve Eq. (4.30) and subsequently Eq. (4.31). The results are illustrated in Figs. 4.12 and 4.13. The photon wavepackets $\varphi^{-1}\Omega_1(t, L)$ of the coherently scattered x-ray from target 1 only are showed in Figs. 4.12a and 4.13a, c, and e. The resulting NFS time spectra $|\varphi^{-1}\Omega_2(t, L)|^2$ registered by the detector in Fig. 4.11 are showed in Figs. 4.12b and 4.13b, d, and f. The presence of two targets of $\xi_1 = \xi_2 = 1$ under $\Delta_{B1} = \Delta_{B2} = 30\Gamma$ results in a faster coherent decay that proceeds with the effective resonant depth of $\xi = 2$, i.e., double the thickness of each target [51]. The magnetic fields $\mathbf{B}_1$ and $\mathbf{B}_2$ are switched off at $T_{\text{off}} = 51$ ns and back on at $T_{\text{on}} = 100$ ns in the original direction. For continuous phase in Fig. 4.12a, the intensity of the scattered field does not change after the retrieval as showed in Fig. 4.12b. If, however, in the case of Fig. 4.13 the phase of the retrieved field is π-modulated by turning on the opposite magnetic field $-\mathbf{B}_1$ as showed in Fig. 4.13a, the detected signal in Fig. 4.13b is significantly suppressed due to destructive inter-

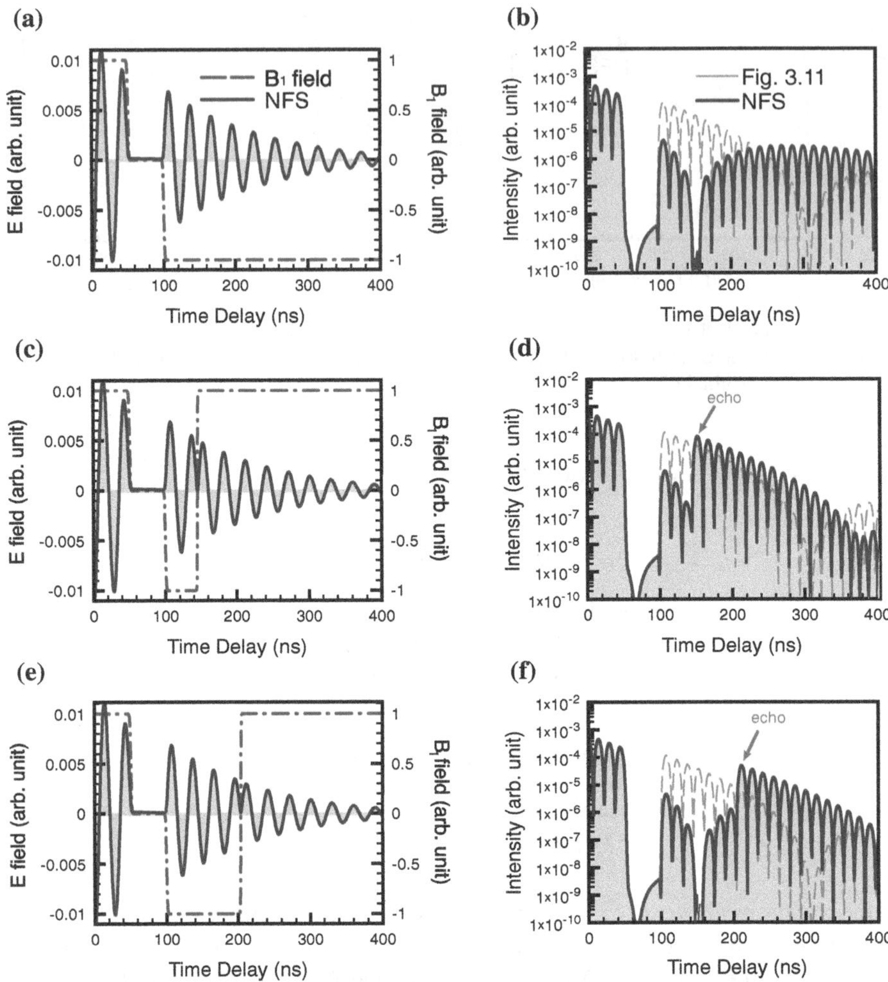

Fig. 4.13 NFS signal of the two-target x-ray interferometer and the mechanics-free x-ray echo. **a**, **c**, **e** illustrate the electric field $\varphi^{-1}\Omega_1(t, L)$ of the x-ray scattered from target 1 only. **b**, **d**, **f** depict the resulting NFS signal $|\varphi^{-1}\Omega_2(t, L)|^2$ registered by the detector shown in Fig. 4.11. The magnetic fields ($\mathbf{B}_1$, $\mathbf{B}_2$) are turned off at 51 ns, and then ($-\mathbf{B}_1$, $\mathbf{B}_2$) are turned on at 100 ns. A second reversal of $\mathbf{B}_1$ orientation is performed at **c**, **d** 145.8 ns and **e**, **f** 204.3 ns and is accompanied by the appearance of the x-ray echo

ference between the two scattering channels. In turn, a second magnetic field rotation back at a node value $E^{(1)}(t > 100\text{ ns}) = 0$ changes the phase of the single photon again with π. This further π phase modulation is explicitly seen in Fig. 4.13b and d. An echo is produced because of constructive interference as it can be seen in Fig. 4.13 for the rotation of $\mathbf{B}_1$ back at (d) $t = 145.8$ ns and (f) $t = 204.3$ ns.

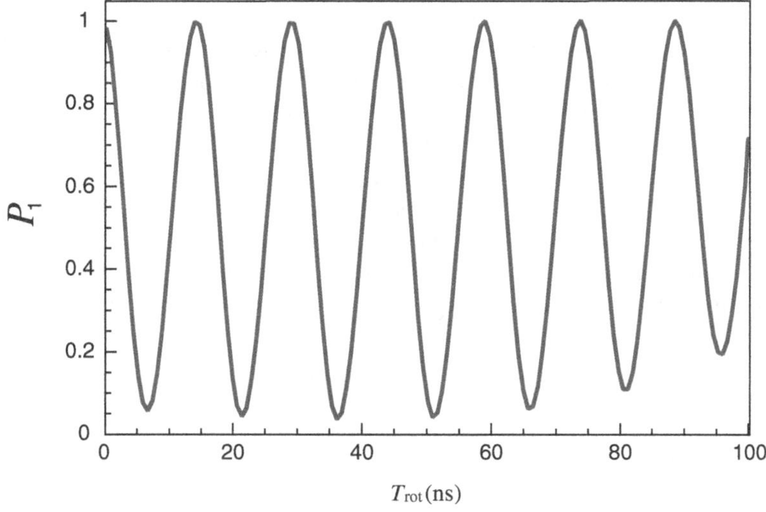

Fig. 4.14 The interference fringe from the two-target hard x-ray interferometer

We emphasize that photon storage is not necessary for phase modulation, since the echo effect clearly shows that the phase control is only achieved via the manipulation of nuclear dynamics, i.e., the time reversal effect [55, 56]. It becomes apparent that without turning off the hyperfine magnetic fields the magnetically induced nuclear exciton echo itself also provides a simple way of photon storage: inverse magnetic fields in target 1 and 2 result in a significant suppression of the scattered x-ray light. To show this effect in Fig. 4.14, we numerically solve the case of Fig. 4.13a and b without any photon storage (no turning off $\mathbf{B}_1$ and $\mathbf{B}_2$) and plot the parameter:

$$P_1(T_{\text{rot}}) = \frac{\int_{T_{\text{rot}}}^{T_{\text{rot}}+1/\Gamma} |\Omega_2^p(t, L)|^2 dt}{\int_{T_{\text{rot}}}^{T_{\text{rot}}+1/\Gamma} |\Omega_2^{up}(t, L)|^2 dt}, \tag{4.33}$$

where $|\Omega_2^{up(p)}(t, L)|^2$ is the NFS signal with (without) reversing $\mathbf{B}_1$, and T_{rot} is the time point of reversing $\mathbf{B}_1$. The red solid line in Fig. 4.14 shows the interference fringe, where each node is corresponding to the minimum of the quantum beat and results from the total destructive interference between channel (2) and (3) in Fig. 4.11. The raising bottom nodes are caused by the growing contribution of channel (4). On the other hand, the oscillation pattern of the interference fringe proves one can continuously tune the relative phase between two scattering channels simply by reversing $\mathbf{B}_1$ at different times T_{rot}. Also, the minima of Fig. 4.14 correspond to the NFS signal suppression due to the destructive interference between channel (2) and (3) in Fig. 4.11, and thus show that the magnetically induced nuclear exciton echo itself provides another convenient solution for photon storage. A sequence of two 180° rotations of the magnetic field direction in target 1 at the quantum beat minima can lead to storage and retrieval of the x-ray photon π phase-modulated. This can be experimentally achieved in antiferromagnets as $^{57}FeBO_3$ with strong intrinsic hyperfine magnetic fields that can be rotated with the help of a weak 10 G external field [1]. Fast 180° magnetic field rotations in such materials have been demonstrated [55]. This specific case of magnetic switching in a two-target setup preserves the photon polarization and can modulate the photonic phase but is less robust compared to the scheme presented in Sect. 4.2.1, since both efficiency of the storage and the phase of the released photon depend on the rotation moment. Nevertheless, the magnetically induced nuclear exciton echo might provide an additional experimentally accessible setup to investigate mechanical-free x-ray storage and phase modulation of a single-photon wave packet.

4.4 Summary

In summary, we have put forward the possibilities of coherent storage, π and continuous phase modulation for single hard x-ray photon wavepacket in a NFS setup with an ^{57}Fe enriched solid-state target. Our coherent storage scheme demonstrated in Sect. 4.2 is mainly based on the quantum destructive interference between two nuclear transition currents associated to the $|\frac{1}{2}, \pm\frac{1}{2}\rangle \rightarrow |\frac{3}{2}, \pm\frac{1}{2}\rangle$ nuclear transitions in a ^{57}Fe nucleus. For triggering and using this quantum interference, a phase evolution of two nuclear currents (or coherences) is introduced by applying a hyperfine magnetic field. In experiments, the so-called quantum beat [48] in the measured NFS time spectra is directly associated to this magnetically induced phase evolution of the nuclear transition currents. From the pattern of quantum beat in the NFS time spectrum, one could realize that the destructive interference occurs regularly, and the beating frequency is proportional to the strength of the applied magnetic field. Based on this observation, we have suggested using the regularly destructive interference property of the quantum beat and turning off the hyperfine magnetic field at each time point when a minimum of quantum beat signal occurs. This operation freeze the nuclei in a special excited state such that the emission of radiation is suppressed, and the single hard x-ray photon is stored in this special nuclear state during the magnetic-field-free period. Due to the fact that the interaction Hamiltonian also vanishes during the storage time, the properties (polarization, phase, etc) of the stored photon are maintained as well. Later on, by turning on the magnetic field, the stored single photon will be retrieved. Because the photonic property is conserved in the whole access process, we term the operation as coherent storage. We emphasize that instead of coherent storage, the scheme achieved by Ref. [1] is an "incoherent" storage, reminding that the interaction Hamiltonian does not vanish and alters the property of the stored single photon during the storage period. On the other hand, our phase modulation scheme presented in Sect. 4.2 is based on the time reversal of the above mentioned magnetically induced phase evolution of the nuclear transition currents. Additionally, the "phase" of a single hard x-ray photon wavepacket is clearly defined by a measurement using a two-target interferometer illustrated in Sect. 4.3. By reversing direction of the hyperfine magnetic field applied to one of the two targets, the time-reversed evolution of the nuclear currents in the modulated target will introduce a relative phase shift between two scattering channels of single photons. This phase modulation will result in a NFS signal suppression or echo which can be treated as the experimental signature of implementing the proposal demonstrated in Sect. 4.3.

References

1. Y.V. Shvyd'ko, T. Hertrich, U. van Bürck, E. Gerdau, O. Leupold, J. Metge, H.D. Rüter, S. Schwendy, G.V. Smirnov, W. Potzel, P. Schindelmann, Storage of nuclear excitation energy through magnetic switching. Phys. Rev. Lett. **77**, 3232 (1996)
2. M.D. Lukin, Colloquium : Trapping and manipulating photon states in atomic ensembles. Rev. Mod. Phys. **75**, 457 (2003)
3. G. Moore, Cramming more components onto integrated circuits. Proc. IEEE **86**, 82 (1998)
4. A. Politi, M.J. Cryan, J.G. Rarity, S. Yu, J.L. O'Brien, Silica-on-silicon waveguide quantum circuits. Science **320**, 646 (2008)
5. H. Specht, J. Bochmann, M. Mücke, B. Weber, E. Figueroa, D. Moehring, G. Rempe, Phase shaping of single-photon wave packets. Nat. Photonics **3**, 469 (2009)
6. Y. Shvyd'ko, *X-Ray Optics: High-Energy-Resolution Applications* (Springer, Berlin, 2004)
7. Y. Shvyd'ko, S. Stoupin, A. Cunsolo, A. Said, X. Huang, High-reflectivity high-resolution x-ray crystal optics with diamonds. Nat. Phys. **6**, 196 (2010)
8. Y. Shvyd'Ko, S. Stoupin, V. Blank, S. Terentyev, Near-100% bragg reflectivity of x-rays. Nat. Photonics **5**, 539 (2011)
9. S. Chen, H. Wu, Y. Chang, Y. Lee, W. Sun, S. Chang, Y. Stetsko, M. Tang, M. Yabashi, T. Ishikawa, Coherent trapping of x-ray photons in crystal cavities in the picosecond regime. Appl. Phys. Lett. **93**, 141105 (2008)
10. F. Pfeiffer, C. David, M. Burghammer, C. Riekel, T. Salditt, Two-dimensional x-ray waveguides and point sources. Science **297**, 230 (2002)
11. A. Jarre, C. Fuhse, C. Ollinger, J. Seeger, R. Tucoulou, T. Salditt, Two-dimensional hard x-ray beam compression by combined focusing and waveguide optics. Phys. Rev. Lett. **94**, 074801 (2005)
12. K. Liss, R. Hock, M. Gomm, B. Waibel, A. Magerl, M. Krisch, R. Tucoulou, Storage of x-ray photons in a crystal resonator. Nature **404**, 371 (2000)
13. Y.V. Shvyd'ko, M. Lerche, H.-C. Wille, E. Gerdau, M. Lucht, H.D. Rüter, E.E. Alp, R. Khachatryan, X-ray interferometry with microelectronvolt resolution. Phys. Rev. Lett. **90**, 013904 (2003)
14. S.-L. Chang, Y.P. Stetsko, M.-T. Tang, Y.-R. Lee, W.-H. Sun, M. Yabashi, T. Ishikawa, X-ray resonance in crystal cavities: Realization of fabry-perot resonator for hard x rays. Phys. Rev. Lett. **94**, 174801 (2005)
15. C. Buth, R. Santra, L. Young, Electromagnetically induced transparency for x rays. Phys. Rev. Lett. **98**, 253001 (2007)
16. T.E. Glover, M.P. Hertlein, S.H. Southworth, T.K. Allison, J. van Tilborg, E.P. Kanter, B. Krässig, H.R. Varma, B. Rude, R. Santra, A. Belkacem, L. Young, Controlling x-rays with light. Nat. Phys. **6**, 69 (2010)
17. B. Dromey, D. Adams, R. Hörlein, Y. Nomura, S. Rykovanov, D. Carroll, P. Foster, S. Kar, K. Markey, P. McKenna et al., Diffraction-limited performance and focusing of high harmonics from relativistic plasmas. Nat. Phys. **5**, 146 (2009)

18. S. Shwartz, S.E. Harris, Polarization entangled photons at x-ray energies. Phys. Rev. Lett. **106**, 080501 (2011)
19. S. Shwartz, R.N. Coffee, J.M. Feldkamp, Y. Feng, J.B. Hastings, G.Y. Yin, S.E. Harris, X-ray parametric down-conversion in the langevin regime. Phys. Rev. Lett. **109**, 013602 (2012)
20. E.P. Kanter, B. Krässig, Y. Li, A.M. March, P. Ho, N. Rohringer, R. Santra, S.H. Southworth, L.F. DiMauro, G. Doumy, C.A. Roedig, N. Berrah, L. Fang, M. Hoener, P.H. Bucksbaum, S. Ghimire, D.A. Reis, J.D. Bozek, C. Bostedt, M. Messerschmidt, L. Young, Unveiling and driving hidden resonances with high-fluence, high-intensity x-ray pulses. Phys. Rev. Lett. **107**, 233001 (2011)
21. N. Rohringer, D. Ryan, R. London, M. Purvis, F. Albert, J. Dunn, J. Bozek, C. Bostedt, A. Graf, R. Hill et al., Atomic inner-shell x-ray laser at 1.46 nanometres pumped by an x-ray free-electron laser. Nature **481**, 488 (2012)
22. O. Kocharovskaya, R. Kolesov, Y. Rostovtsev, Coherent optical control of Mössbauer spectra. Phys. Rev. Lett. **82**, 3593 (1999)
23. R. Coussement, Y. Rostovtsev, J. Odeurs, G. Neyens, H. Muramatsu, S. Gheysen, R. Callens, K. Vyvey, G. Kozyreff, P. Mandel, R. Shakhmuratov, O. Kocharovskaya, Controlling absorption of gamma radiation via nuclear level anticrossing. Phys. Rev. Lett. **89**, 107601 (2002)
24. T.J. Bürvenich, J. Evers, C.H. Keitel, Nuclear quantum optics with x-ray laser pulses. Phys. Rev. Lett. **96**, 142501 (2006)
25. W.-T. Liao, A. Pálffy, C.H. Keitel, Coherent storage and phase modulation of single hard-x-ray photons using nuclear excitons. Phys. Rev. Lett. **109**, 197403 (2012)
26. W.-T. Liao, A. Pálffy, C.H. Keitel, Nuclear coherent population transfer with x-ray laser pulses. Phys. Lett. B **705**, 134 (2011)
27. J. Arthur et al., *Linac Coherent Light Source (LCLS) Conceptual Design Report* (SLAC, Stanford, 2002)
28. Xfel@sacla, http://xfel.riken.jp/eng/sacla/ (2012)
29. M. Altarelli et al., *XFEL: The European X-Ray Free-Electron Laser, Technical Design Report* (DESY, Hamburg, 2009)
30. U.v. Bürck, Coherent pulse propagation through resonant media. Hyperfine Interact. **123/124**, 483 (1999)
31. Y.V. Shvyd'ko, Motif: Evaluation of time spectra for nuclear forward scattering. Hyperfine Interact. **125**, 173 (2000)
32. R. Röhlsberger, *Nuclear Condensed Matter Physics with Synchrotron Radiation: Basic Principles, Methodology and Applications* (Springer, Berlin, 2004)
33. R. Röhlsberger, K. Schlage, B. Sahoo, S. Couet, R. Rüffer, Collective lamb shift in single-photon superradiance. Science **328**, 1248 (2010)
34. R. Röhlsberger, H. Wille, K. Schlage, B. Sahoo, Electromagnetically induced transparency with resonant nuclei in a cavity. Nature **482**, 199 (2012)
35. A. Pálffy, C.H. Keitel, J. Evers, Single-photon entanglement in the keV regime via coherent control of nuclear forward scattering. Phys. Rev. Lett. **103**, 017401 (2009)
36. S.L. Ruby, Mössbauer experiments without conventional sources. J. Phys. Colloques **35**, C6–209 (1974)
37. G.K. Shenoy, Scientific legacy of Stanley Ruby, in *NASSAU 2006* (Springer, Berlin, 2007), p. 5
38. G.V. Smirnov, U. van Bürck, W. Potzel, P. Schindelmann, S.L. Popov, E. Gerdau, Y.V. Shvyd'ko, H.D. Rüter, O. Leupold, Propagation of nuclear polaritons through a two-target system: Effect of inversion of targets. Phys. Rev. A **71**, 023804 (2005)
39. M.O. Scully, E.S. Fry, C.H.R. Ooi, K. Wódkiewicz, Directed spontaneous emission from an extended ensemble of n atoms: Timing is everything. Phys. Rev. Lett. **96**, 010501 (2006)
40. M. Scully, Correlated spontaneous emission on the Volga. Laser Phys. **17**, 635 (2007)
41. M.O. Scully, M.S. Zubairy, *Quantum Optics* (Cambridge University Press, Cambridge, 1997)
42. A.A. Svidzinsky, J.-T. Chang, M.O. Scully, Dynamical evolution of correlated spontaneous emission of a single photon from a uniformly excited cloud of *n* atoms. Phys. Rev. Lett. **100**, 160504 (2008)

43. E. Sete, A. Svidzinsky, H. Eleuch, Z. Yang, R. Nevels, M. Scully, Correlated spontaneous emission on the danube. J. Mod. Opt. **57**, 1311 (2010)
44. M.D. Crisp, Propagation of small-area pulses of coherent light through a resonant medium. Phys. Rev. A **1**, 1604 (1970)
45. Y. Shvyd'ko, Nuclear resonant forward scattering of x rays: Time and space picture. Phys. Rev. B **59**, 9132 (1999)
46. A. Pálffy, J. Evers, C.H. Keitel, Electric-dipole-forbidden nuclear transitions driven by super-intense laser fields. Phys. Rev. C **77**, 044602 (2008)
47. Y.V. Shvyd'ko, U. van Bürck, W. Potzel, P. Schindelmann, E. Gerdau, O. Leupold, J. Metge, H.D. Rüter, G.V. Smirnov, Hybrid beat in nuclear forward scattering of synchrotron radiation. Phys. Rev. B **57**, 3552 (1998)
48. Y.V. Shvyd'ko, U. van Bürck, Hybrid forms of beat phenomena in nuclear forward scattering of synchrotron radiation. Hyperfine Interact. **123–124**, 511 (1999)
49. R. Röhlsberger, J. Evers, private communications (2012)
50. A. Pálffy, J. Evers, Coherent control of nuclear forward scattering. J. Mod. Opt. **57**, 1993 (2010)
51. G.V. Smirnov, U. van Bürck, J. Arthur, S.L. Popov, A.Q.R. Baron, A.I. Chumakov, S.L. Ruby, W. Potzel, G.S. Brown, Nuclear exciton echo produced by ultrasound in forward scattering of synchrotron radiation. Phys. Rev. Lett. **77**, 183 (1996)
52. H. Jex, A. Ludwig, F.J. Hartmann, E. Gerdau, O. Leupold, Ultrasound-induced echoes in the nuclear-exciton decay utilising synchrotron radiation bragg-scattered by a quartz crystal. Europhys. Lett. **40**, 317 (1997)
53. N. Miura, T. Osada, S. Takeyama, Research in super-high pulsed magnetic fields at the mega-gauss laboratory of the university of tokyo. J. Low Temp. Phys. **133**, 139 (2003)
54. R. Röhlsberger, T.S. Toellner, W. Sturhahn, K.W. Quast, E.E. Alp, A. Bernhard, E. Burkel, O. Leupold, E. Gerdau, Coherent resonant x-ray scattering from a rotating medium. Phys. Rev. Lett. **84**, 1007 (2000)
55. Y. Shvyd'ko, Perturbed nuclear scattering of synchrotron radiation. Hyperfine Interact. **90**, 287 (1994)
56. Y. Shvyd'ko, T. Hertrich, J. Metge, O. Leupold, E. Gerdau, H. Rüter, Reversed time in möss-bauer time spectra. Phys. Rev. B **52**, R711 (1995)
57. Y. Hasegawa, Y. Yoda, K. Izumi, T. Ishikawa, S. Kikuta, X.W. Zhang, M. Ando, Phase transfer in time-delayed interferometry with nuclear resonant scattering. Phys. Rev. Lett. **75**, 2216 (1995)
58. Y. Hasegawa, S. Kikuta, Time-delayed interferometry with nuclear resonant scattering. Hyperfine Interact. **123–124**, 721 (1999)
59. P. Helistö, E. Ikonen, T. Katila, K. Riski, Coherent transient effects in Mössbauer spectroscopy. Phys. Rev. Lett. **49**, 1209 (1982)
60. P. Helistö, I. Tittonen, M. Lippmaa, T. Katila, Gamma echo. Phys. Rev. Lett. **66**, 2037 (1991)

Chapter 5
Coherence Enhanced Optical Determination of the ^{229}Th Isomeric Transition

In this chapter, we will join our so far gained knowledge on light-nuclei interaction and collective effects to a developing community, the ^{229}*Thorium community*. The ^{229}Th nucleus has a low-level isomeric transition which occupies a very special position in the whole known isotope chart. Looking down on the landscape of "nuclear transition energy to linewidth ratio", the ^{229}Th nucleus is expected to stand on the top of the Mount Everest with an altitude of $E/\Delta E \approx 10^{20}$. Inspired by this sharp peak, the idea of a nuclear clock [1] based on this isomeric transition was proposed by E. Peik and C. Tamm at the Physikalisch-Technische Bundesanstalt (PTB, Germany) in 2003. This may become the next generation nuclear time standard, and provide many potential applications for metrology and cosmology, e.g., to study the time variation of physical constants. In addition to metrology, the nuclear clock transition provides a remarkable platform for developing nuclear quantum optics [2], studying the interplay between nuclei and its outer electrons [3], investigating quantum information [4] and building a γ-ray laser [5]. However, the basic problem in the thorium community is that the precise value of the isomeric transition energy is still unknown.

The central issue of this chapter is trying to offer a solution to the basic problem of finding the ^{229m}Th isomeric energy. We start with an introduction to the ^{229}Th nucleus. Section 5.1 is devoted to give a flavor about how difficult it is to measure the ^{229m}Th energy from a historical perspective. Also, many modern approaches are briefly introduced. Out of these modern schemes, the advantage of a solid-state thorium doped crystal will be discussed in Sect. 5.2. Motivated by these advantages, we discuss and provide solutions based on an electromagnetically modified nuclear forward scattering setup (two-field setup) to the obstacles of the experiments in Sect. 5.2. In Sect. 5.3, we first analyze the time scale of the considered system, and subsequently the numerical results are demonstrated. Finally, we find the best setup of our proposals is a two-field setup reminding of electromagnetically induced transparency (EIT) [6]. This scheme makes it possible to directly measure the ^{229m}Th isomeric energy with a very low uncertainty of only 10 Hz.

W.-T. Liao, *Coherent Control of Nuclei and X-Rays*, Springer Theses,
DOI: 10.1007/978-3-319-02120-1_5,

5.1 Introduction of the ^{229}Thorium Nucleus

This is a journey over 35 years about how physicists have tried to pinpoint one optical nuclear transition of ^{229}Thorium. Although we are still in a tunnel of darkness, the light at the end seems to approach.

Thorium (proton number 90) has six naturally occurring members in its isotope family, however, none of them stable. The most abundant thorium isotope on the Earth is ^{232}Th due to its very long half-life of 14.05 billion years. Other isotopes are ^{228}Th, ^{229}Th, ^{230}Th, ^{231}Th and ^{234}Th. The main character of this chapter is ^{229}Th. ^{229}Th is mainly produced by the decay of ^{233}U and subsequently disintegrates into its daughter nucleus ^{225}Ra via α-decay [7]. ^{229}Th can also be the daughter of ^{229}Ac and ^{229}Pa via β-decay [8]. The special feature of ^{229}Th nucleus is that there is an isomeric first excited state ^{229m}Th featuring a rarely low energy of 7.8 $\pm$ 0.5 eV [9, 10]. This low nuclear excitation corresponding to the vacuum ultraviolet (VUV) region with a wavelength of 163 $\pm$ 11 nm. The half-life of ground state ^{229g}Th and that of isomer state ^{229m}Th are 7880 $\pm$ 120 years [11] and 6 $\pm$ 1 h [4], respectively. The remarkably low energy and very long lifetime of ^{229m}Th lead to a very sharp energy to linewidth ratio of ^{229m}Th, i.e., $E/\Delta E \approx 10^{20}$. This considered high ratio gives the tempting reason to build a so-called nuclear clock [1] which may provide the next generation time standard. However, the main impediment is the so far elusive exact transition frequency. Several contradicting measured values of ^{229m}Th energy exist in the literature [7, 9, 10, 12–14], and so far no direct conclusive observation of this isomeric transition was reported.

The thorium community has its elaborate Bildungsroman, and here we would like to give a short overview of the history and the present state of the art on the isomeric transition. The story of ^{229}Thorium nucleus kicked off in 1976. L. A. Kroger and C. W. Reich [12] performed the prelude when they studied the ^{229}Th nuclear structure with γ-ray spectroscopy and suggested the existence of a $K^{\pi} = \frac{3}{2}^{+}$ rotational band to explain their observation. At that time the data showed no observable energy separation between different transitions from excited states to the ground states of $K^{\pi} = \frac{5}{2}^{+}$ and $\frac{3}{2}^{+}$ band. The authors of [12] further checked the γ-ray peaks likely containing decays to both the $\frac{5}{2}^{+}$ and $\frac{3}{2}^{+}$ levels and then finally concluded that a slightly spaced doublet structure might exist within an energy separation of less than 0.1 keV. In 1990, D. G. Burke et al. provided an observation which supported the existence of the proposed $K^{\pi} = \frac{3}{2}^{+}$ band in ^{229}Th. They compared the spectroscopic strengths for the rotational band members in ^{231}Th and ^{229}Th [15] and found the distributions of strengths for the proposed $K^{\pi} = \frac{5}{2}^{+}$ and $K^{\pi} = \frac{3}{2}^{+}$ rotational bands in ^{229}Th are virtually identical to those of ^{231}Th. This identity seemed to confirm Kroger and Reich's idea. On the other hand, they concluded the isomeric energy might be less than 5 eV.

In 1990, C. W. Reich and R. G. Helmer studied the energies of selected γ-rays subsequently emitted after an α-decay of ^{233}U and obtained a ^{229m}Th energy of 1 $\pm$ 4 eV [7]. Later on, they tried to improve the energy resolution of the used detector down to 300–900 eV and obtained an isomeric energy of 3.5 ± 1 eV in 1994 [13]. This result became the most accepted value for a long time. G. M. Irwin and K. H. Kim claimed in 1997 that they directly observed the ultraviolet photon from the deexcitation of the ^{229m}Th with two different solid ^{233}U samples [16]. A similar conclusion was also claimed by D. S. Richardson et al. with a liquid sample in 1998 [17]. Their observed value for the isomeric transition were 3.5 eV [16] and likely 4 eV [17], respectively. However, in 1999, the studies of three different groups indicated that these conclusions must be based on the incorrect interpretation of experimental data [18–20]. G. M. Irwin and K. H. Kim later agreed that their observation might result from the α-particle induced fluorescence of air, actually N_2, surrounding the sample [18, 19, 21]. K. Gulda et al. studied the nuclear structure of ^{229}Th measuring the γ-rays following the β-decay of ^{229}Ac in 2002 [8]. They again identified the $K^\pi = \frac{5}{2}^+$ and $K^\pi = \frac{3}{2}^+$ rotational bands, but could not conclude on the ^{229m}Th energy. V. Barci et al. estimated the ^{229m}Th energy at about 3.4 ± 1.8 eV from their experiment with ^{233}U samples in 2003 [22]. In 2005, Z. O. Guimarães-Filho and O. Helene reexamined these results together with [7, 13] and estimated the ^{229m}Th energy at 5.5 ± 1 eV [23].

In 2007, B. R. Beck et al. demonstrated a new result 7.6 ± 0.5 eV with a much better detector resolution of 30 eV [9]. They used the NASA x-ray spectrometer (XRS) to measure the γ-ray decay of the 71.82 keV state of ^{229}Th nucleus. Subsequently in 2009 this value was corrected to 7.8 ± 0.5 eV because of two missing 42.43 keV $\rightarrow$ ^{229m}Th and 29.19 keV $\rightarrow$ ^{229g}Th branching ratios [10]. For more review and details, one can refer to S. L. Sakharov's analysis in 2010 [14]. The value 7.8 eV may explain why the ^{229m}Th deexcitation could escape from all observations before, as physicists were looking for it in a possibly wrong detection window! Because of the nice idea of the nuclear clock [1] together with the very compelling experiment of observing 7.8 eV [9, 10], the thorium community started to develop. Several groups worldwide pursue today the "Holy Grail" of the isomeric transition.

In 2008, C. J. Campbell et al. at Georgia Institute of Technology (USA) firstly produced laser-cooled $^{232}Th^{3+}$ crystals in a linear rf Paul trap [24]. They also achieved so-called Wigner crystals of $^{229}Th^{3+}$ in 2011, and recently proposed the idea of single $^{229}Th^{3+}$ nuclear clock [25]. In 2009, T. T. Inamura and H. Haba at RIKEN Nishina Center (Japan) studied the isomeric energy utilizing a ^{229}Th loaded hollow cathode with the use of nuclear excitation by electron transition (NEET) [26]. They concluded the ^{229m}Th appears to lie in 3 eV $\sim$ 7 eV region [27]. Two groups at Lawrence Livermore National Laboratory (USA) and Los Alamos National Laboratory (USA) tried to measure the ^{229m}Th lifetime by collecting ^{229}Th recoils from the alpha decay of ^{233}U into MgF_2 plates and measuring the subsequent UV photon emission. Very recently, X. Zhao et al. were the first who directly observed the isomeric decay and concluded the isomeric lifetime is 6 ± 1 h. Interestingly, no internal conversion electrons were observed in their experiment [4]. E. Peik's group at PTB (Germany) is trying to excite the isomeric transition via the electron bridge effect in Th^+ ions [3]. At the Max Planck Institute of Quantum Optics (Germany), L. v.d. Wense et al. developed an apparatus composed of a 'MLL IonCatcher' and an UV focusing system [28] to directly identify the isomeric transition. Another option of identifying ^{229m}Th is to measure its hyperfine structure [29]. This project is pursued at IGISOL in Jyväskylä (Finland) together with University of Mainz (Germany).

Finally, a very promising approach using a 10^{18} ^{229}Th /cm^3 doped VUV transparent solid-state crystal termed as ^{229}Th:Crystal is investigated by two groups, E. Hudson's group at University of California Los Angeles (USA) [30, 31] and T. Schumm's group at TU Vienna (Austria) [32]. Their ultimate goal is to build solid-state nuclear clocks. Due to the advantage of the very high ^{229}Th concentration, we consider the ^{229}Th:Crystal approach for our theoretical investigation in this chapter.

5.2 Forward Detection Solves Some Critical Problems

So far, 7.8 ± 0.5 eV, the adopted value of energy splitting of the ^{229}Th ground-state doublet, was indirectly measured by B. R. Beck et al. at Lawrence Livermore National Lab [9, 10] in 2007. The one eV uncertainty is too large to let one use ^{229}Th in any future application, hence enormous efforts were made to narrow the uncertainty of measurements as mentioned in the introduction. In contrast to other proposals, the solid state approach uses samples containing a very high nuclear concentration of 10^{18}/cm^3 [30–32]. This huge number of nuclei significantly increases the absorption/emission rate of VUV photons and allows one to use synchrotron radiation or weak VUV lasers for Mössbauer spectroscopy. Thus, in this chapter we will focus on investigating the optical nuclear transition in ^{229}Th doped VUV transparent solid state crystals.

Up to the date of writing, there is no direct observation of ^{229}Th isomeric transition via Mössbauer spectroscopy with solid state nuclear samples. The only optical experiments were performed by E. Hudson's group [30, 31] with doped ^{232}Th in high-energy band gap crystals. Figure 5.1a illustrates the fluorescence setup used to identify suitable host crystals. A ^{232}Th:Crystal is illuminated by the synchrotron radiation from the Berkeley Advance Light Source with a fixed photon energy for 5 seconds. Afterwards the fluorescence from the crystals is collected at 90 degree over the next 10.7 seconds. The photon energy is scanned from 7 eV to 11 eV covering the expected 7.8(5) eV isomeric transition energy. Various crystals produce different fluorescence energy spectra demonstrating various degree of transparency to VUV photons. In particular, lattice defects such as color centers affect the transparency. The authors of Ref. [30] concluded the most suitable crystal is Th:$LiCaAlF_6$ out of the analyzed crystals: Na_2ThF_6, Th:$LiCaAlF_6$, Th:YLF and Th:NaYF. Another promising host crystal is CaF_2 used by T. Schumm's group [32]. However, the question arises: does the measured fluorescence really reflect solely the response of ^{229}Th nuclear excitation in real runs[1]?

---

[1] By using thorium ions doped crystals for the current purpose, one has to exclude all kinds of VUV absorption except nuclear response, e.g., the band gap of the host crystal should be larger than 7.8 eV; this is the purpose of [30, 31].

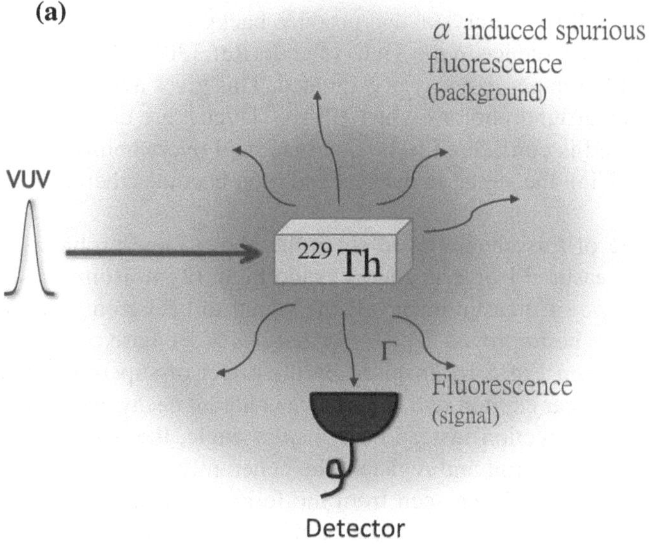

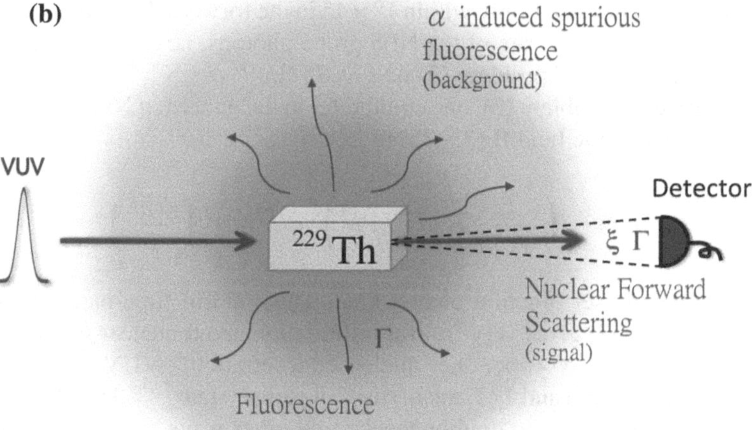

Fig. 5.1 **a** Setup for measuring the decay fluorescence from a ^{229}Th:Crystal [30]. The photon detector used to collect the signal is placed in the perpendicular direction to the propagation of VUV pulse. **b** Setup for measuring time spectra of nuclear forward scattering with a ^{229}Th:Crystal [2]; the photon detector is placed in the forward direction. The *blue* VUV pulse impinges on a Thorium doped crystal and partially excites the nuclei to the isomer state. Isomeric population will experience two decay channels: (i) the spontaneous decay emits fluorescence signal spatially distributed in 4π; (ii) the coherent channel also called nuclear forward scattering or single-photon superradiance emits the signal photon in the forward direction. The *red* aureola depicts the background photons induced by radioactive α-decay of the ground state Thorium nuclei

In particular, doped ^{229}Th is radioactive and disintegrate to ^{225}Ac via α-decay. Emitted α particles will hit the crystal and produce background photons labeled as α induced spurious fluorescence in the Th:Crystal. In Ref. [30, 31] a producing rate of 0.3 background photon per α-decay is estimated. This gives a total counting rate of few MHz for background photons when 10^{18} ^{229}Th/cm^3 are doped in a crystal. This large background is considered to be dominant and overwhelms the scattered VUV signal. Optimizing the signal to background ratio becomes therefore a crucial task.

The central issue of this chapter is to solve the above three problems: (1) the uncertainty of the measured isomeric energy is too high. (2) an isomeric signature is desired for any type of measurement. (3) the signal to background ratio is low and needs therefore to be improved. The setup that lies at the base of our proposed solutions is illustrated in Fig. 5.1b. Inspired by the effect of superradiant forward emission,[2] i.e., the nuclear exciton will experience coherent decay and emit a single photon in the forward direction instead of 4 π solid angle, the detector is simply moved to the forward direction and registers the coherently scattered photon. One benefit of this setup can easily be seen from the following considerations. Let us consider the case of a ^{229}Th:Crystal with a size of $3 \times 3 \times 10$ mm and with a ^{229}Th concentration of 10^{18}/cm^3. The half lifetime of ^{229}Th is 7880 yr and we consider a 0.3 background photon per α-decay [30]. The total counting rate of background photons is then 0.75 MHz, which could be significantly suppressed down to 1.8 Hz by registering coherent signals only within $1° \times 1°$ in the forward direction. Thus, 20 Hz would suffice as counting rate of the NFS probe photons, i.e., a signal-background ratio greater than 10, and problem (3) is solved [2].

To overcome the problem (2), one could refer to the pattern of NFS time spectrum in an external magnetic field **B** [33–35] in free space[3]:

$$I(t) \sim \left(\frac{\xi}{\sqrt{\xi \Gamma t}} J_1 \left[2\sqrt{\xi \Gamma t} \right] \right)^2 e^{-\Gamma t} cos^2 \left(\frac{\Delta_B t}{2} \right) \tag{5.1}$$

where J_1 is the Bessel function of first kind. This kind of time-based Mössbauer spectroscopy [36, 37] gives very fruitful information about nuclear properties, e.g., the resonant depth $\xi \sim N \times \sigma \times L$, i.e., nuclear concentration × VUV absorption cross section × sample length and $1/\Gamma$ the lifetime of isomer state ^{229m}Th. All parameters in Eq. (5.1) can be precisely double-checked by other ways [4, 30, 31], therefore one can plug these pre-measured factors in Eq. (5.1) to examine if the observed NFS time spectra solely reflect the isomeric response. Furthermore, the quantum beat induced by the Zeeman splitting $\Delta_B \sim (\mu_g + \mu_m) \cdot \mathbf{B}$ offers the possibility of directly measuring the isomeric magnetic dipole moment μ_m if the strength of external magnetic field **B** is well controlled, and the ground state magnetic dipole

[2] The detailed discussion of single photon superradiance is presented in Sect. 4.1.2.

[3] "in free space" means we only consider the spontaneous γ-decay as the decoherence process. However, in crystal environments the spin-spin relaxation will come into play and cause an additional decoherence process [32] with rates on the kHz scale. We will consider this effect in the next section.

moment μ_g is determined by nuclear magnetic resonance (NMR) technique [32]. Hence one can gain the certainty that the NFS time spectra are the pure response of ^{229m}Th and provide an unmistakable signature for the nuclear excitation.

With the above discussion we have shown that the problems (2) and (3) are simultaneously solved by detecting the coherent signal in the forward direction. In the next section we will present a new solution to the large uncertainty of problem (1) by applying an additional VUV laser or a static magnetic field to modify the dispersion relation of the isomeric transition of interest.

5.3 Optical Determination of the ^{229}Th Isomeric Transition

5.3.1 Preparing A Simple Initial Nuclear State and the NFS Based Measurements

The internal structure of the ^{229}Th ground state doublet is quite complicated. The nuclear spins for the two rotational band heads are $\frac{3}{2}$ and $\frac{5}{2}$, so there are 10 Zeeman sub-levels and 12 drivable transitions involved in the problem. To make things easier, strategies need to be developed. In 2011 E. V. Tkalya proposed a nuclear γ-ray laser scheme using a gain medium of ^{229}Th:LiCaAlF$_6$ crystal [5]. Here we follow his suggestion of treating nuclear Zeeman sub-levels with quadrupole splitting in our investigation. Figure 5.2 illustrates the Zeeman sub-levels of ^{229}Th nuclei doped in the two crystals of experimental interest. Each state is termed as $|I, m\rangle$ where I is the spin of nuclear state, and m denotes the projection of nuclear spin on the quantization axis. The electric field gradient (EFG) at the nuclear position inside the crystal will introduce the so-called quadrupole energy splitting of approx. 10^{-4} eV in ^{229}Th:LiCaAlF$_6$ [5], and 10^{-7} eV in ^{229}Th:CaF$_2$ [32]. To prepare a simple initial nuclear state, we further adopt the recently proposed ^{229}Th doped crystals under the thermal equilibrium condition [5], i.e., at sub-kelvin temperatures all nuclei equally populate the two ground states. One could take advantage of this simple initial state: an applied laser will drive only one transition in ^{229}Th:CaF$_2$, and maximum two transitions in ^{229}Th:LiCaAlF$_6$, depending on its polarization. Consequently, not only the measured signal but also the mathematical descriptions become easier to handle.

How can the transition frequency of $|\frac{5}{2}, \pm\frac{1}{2}\rangle \leftrightarrow |\frac{3}{2}, \pm\frac{1}{2}\rangle$ in Fig. 5.2a and that of $|\frac{5}{2}, \pm\frac{5}{2}\rangle \leftrightarrow |\frac{3}{2}, \pm\frac{3}{2}\rangle$ in Fig. 5.2b be precisely determined in a nuclear forward scattering setup? For convenience, let us take the setup of Fig. 5.3a as an example (this way works for all of our proposals). A left circularly polarized VUV probe laser Ω_p with a tunable frequency ω_p is applied to drive $|\frac{5}{2}, \frac{5}{2}\rangle \leftrightarrow |\frac{3}{2}, \frac{3}{2}\rangle$ nuclear $\Delta m = -1$ transition. The corresponding transition energy is

$$E_{|\frac{5}{2},\frac{5}{2}\rangle\leftrightarrow|\frac{3}{2},\frac{3}{2}\rangle} = \hbar\left(\Delta_p + \omega_p\right), \tag{5.2}$$

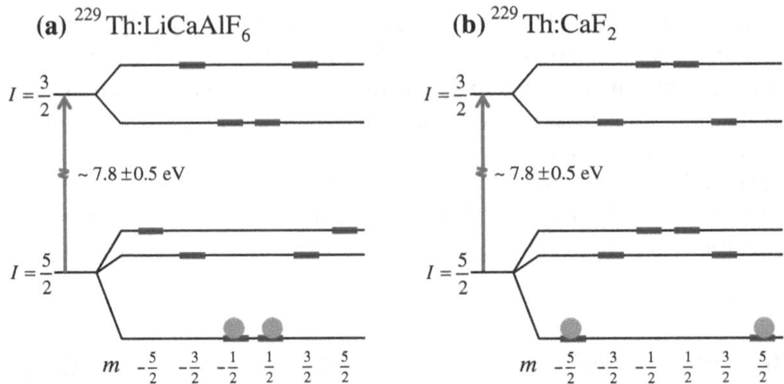

Fig. 5.2 The nuclear quadrupole splitting level scheme at sub-kelvin temperatures in (**a**) ^{229}Th:LiCaAlF$_6$, and in (**b**) ^{229}Th:CaF$_2$

where $\hbar$ denotes the reduced Planck constant, and laser detuning Δ_p is initially unknown. One can determine Δ_p later by fitting the experimental data of NFS time spectra with our theory, and then we eventually could deduce the value of $E_{|\frac{5}{2},\frac{5}{2}\rangle \leftrightarrow |\frac{3}{2},\frac{3}{2}\rangle}$ since ω_p is known in the beginning. Essentially, all transition energies are able to be measured by this NFS-based method utilizing synchrotron radiation or VUV lasers of available ω_p.

5.3.2 *Time Scale, Model, Results and Discussion*

In this section, we will calculate and compare the Δ_p dependent pattern of NFS time spectra using the Maxwell-Bloch equations. Four setups are designed in order to tackle difficulties in two types of doped crystals and also possible obstacles of VUV laser technology.

Let us begin with the time scale analysis. In the ^{229}Th:LiCaAlF$_6$ and ^{229}Th:CaF$_2$ crystals, thorium nuclei at each lattice site interact with their neighboring ions. This complicated environment affects the thorium nuclear structure and causes some decoherence effects. Here we shortly address the decoherence effect named spin-spin relaxation. Recently, G. A. Kazakov et al. discussed the effects of CaF$_2$ crystal lattice environment [32]. They considered the spin-spin interaction $\widehat{H}_{ss}(t)$, i.e., a thorium nucleus interacting with random magnetic fields generated by the spin of its single neighboring fluorine nucleus.[4] This $\widehat{H}_{ss}(t)$ is assumed to be the fastest, weakest random process among the complete Hamiltonian [32] and its time average is $\langle \widehat{H}_{ss}(t) \rangle = 0$. Then the time evolution of the nuclear density matrix reads:

[4] The question might arise: is there any interaction between doping thorium ions? Fortunately, the ions doped in the crystal mostly are ^{229}Th^{4+} therefore all electron spins in its outer shells are paired and produce no net magnetic field [31, 32]. Thus, the interaction between the thorium is negligible.

$$\partial_t \widehat{\rho}(t) = \frac{1}{i\hbar}\left[\widehat{H}(t), \widehat{\rho}(t)\right] - \frac{1}{\hbar^2}\int_{-\infty}^{t}\left\{\overline{\widehat{H}_{ss}(t), [\widehat{H}_{ss}(t'), \rho(t)]}\right\} dt' . \tag{5.3}$$

Here the first term determines the coherent dynamics under the action of the laser, and the second term describes the incoherent spin-spin relaxation. The spin-spin relaxation term can be rewritten as $-\gamma\rho(t)$ with γ being a decay rate of the order of kHz [31, 32]. Eq. (5.3) then becomes

$$\partial_t \widehat{\rho}(t) = \frac{1}{i\hbar}\left[\widehat{H}(t), \widehat{\rho}(t)\right] - \gamma\rho(t) . \tag{5.4}$$

From Eq. (5.4) we can see that $\rho(t) \sim e^{-\gamma t}$ for any coherence, and this limits that the lifetime of any coherent nuclear current $i\rho(t)$ on the order of milliseconds (ms) (γ is much greater than the isomeric spontaneous decay rate). From inspecting the Maxwell-Bloch equation, e.g., Eq. (4.22), we know that the nuclear current $i\rho(t)$ is the source of the emitted coherent photons. The coherent light signal from the thorium doped crystal will therefore also be observed in ms time scale after each incident shot.[5] As discussed in Sect. 4.1.3, to estimate the characteristic time scale of the considered dynamics, the superradiant speed-up effect should also be taken into account. The enhanced coherent decay rate $\xi\Gamma$ (the resonant thickness $\times$ spontaneous decay rate $\sim 10^6 \times 0.1$ mHz) is around 100 Hz such that the dynamical beat may be observed on a 10 ms time scale. However, the spin-spin relaxation ($\gamma > \xi\Gamma$) will wash out the nuclear coherences, leading to the disappearance of the dynamical beat. The single photon superradiance should therefore be observed within the time scale of ms, allowing 100–500 incident shots of the probe pulse per second in a NFS experiment. Together with the required 20 Hz counting rate previously estimated in Sect. 5.2, the probe laser intensity should provide at least 0.04 scattered coherent photon per shot.[6] On the other hand, we can expect the pattern of NFS time spectra to become very different from that of Fig. 4.6 described by Eq. (5.1). Thus, we must consider another shorter-time-scale effect as the isomeric signature, i.e., the quantum beat. We have seen that the quantum beat is caused by the Zeeman effect, therefore we need the help of a hyperfine magnetic field. In addition to the Zeeman splitting, we

[5] This phenomenon can be easily understood with a metaphor. Consider a chorus consisting of singers (A, B, C, ..., Z) and some friends of them (a, b, c, ..., z) at a Christmas party. As long as the chorus members absolutely follow the gestures of their conductor, we can enjoy very nice songs. However, imagine that while singing, chorus member B whispers to his/her friend b for a short time, followed by friend c to member C, F to f, W to w and so on. The performance will be terrible. Comparing the metaphor with our physical system, the laser is the conductor, thorium nuclei are the singers, the nearby friends are the fluorine nuclei, the measured signal is the performed songs, and we can term spin-spin relaxation as whisper-whisper relaxation in the metaphor.

[6] The following situation may help one understand the advantage of the NFS-based method (single photon superradiance). Imagine a PhD student (setting in the 4th year may be more suggestive) who tries to find out the isomeric energy. With the setup of Fig. 5.1a, he must wait 6 h to register one fluorescence photon after an incident pulse. However, the setup of Fig. 5.1b leads the student observe one coherent photon in the forward direction in only few milliseconds. Time is money, and one saves a lot with the NFS-based method.

will propose a two-lasers setup using the Autler-Townes effect [38] in the following. The above analysis provides us with three useful pieces of information

- The characteristic time scale of our system dynamics is ms.
- The NFS-based method significantly shortens the experimental time.
- The quantum beat can provide the isomeric signature.

5.3.2.1 Electromagnetically Modified NFS with One ^{229}Th:CaF$_2$ Crystal

We now discuss how to determine the $|\frac{5}{2}, \frac{5}{2}\rangle \leftrightarrow |\frac{3}{2}, \frac{3}{2}\rangle$ transition energy of the ^{229}Th:CaF$_2$ system depicted in Fig. 5.3. Two cases will be addressed here, the first one showed in (a) using only one probe pulse Ω_p propagating through the crystal, and the second case demonstrated in (b) that uses a second field Ω_c to affect the behavior of the first Ω_p. Since thorium nuclei initially populate only the $|\frac{5}{2}, \pm\frac{5}{2}\rangle$ states in a cooled ^{229}Th:CaF$_2$ crystal, the left circularly polarized probe (blue arrows) will only drive the $|\frac{5}{2}, \frac{5}{2}\rangle \leftrightarrow |\frac{3}{2}, \frac{3}{2}\rangle$ transition. In addition to the VUV probe pulse, another continuous wave VUV laser of linear polarization denoted as couple beam (red wider arrows) is applied. The couple field not only connects the $|\frac{5}{2}, \frac{3}{2}\rangle$ and $|\frac{3}{2}, \frac{3}{2}\rangle$ states but also modifies the dispersion relation of the probe beam.

We study both systems demonstrated in Fig. 5.3a (assuming $\Omega_c = 0$) and Fig. 5.3b by numerically determining the dynamics of the density matrix $\widehat{\rho}(t)$ with the help of the Maxwell-Bloch equations [2, 39, 40]:

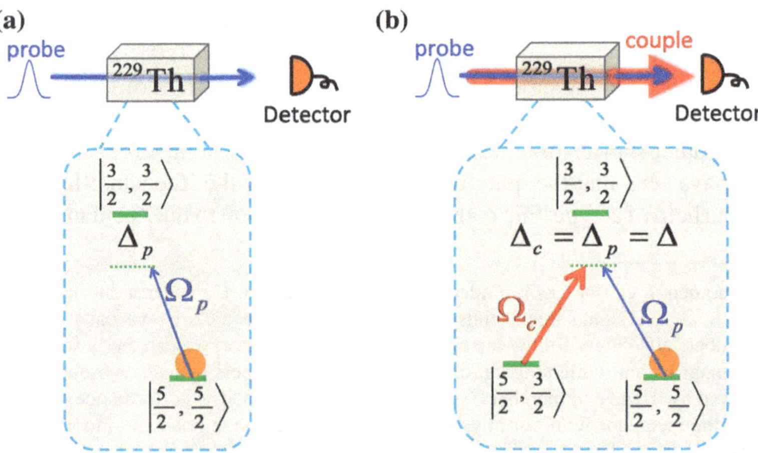

Fig. 5.3 Two cases in ^{229}Th:CaF$_2$. **a** A left circularly polarized probe field drives the $|\frac{5}{2}, \frac{5}{2}\rangle \leftrightarrow |\frac{3}{2}, \frac{3}{2}\rangle$ isomeric transition. **b** The electromagnetically modified forward scattering setup. The red wider arrow denotes the strong couple field while the blue arrow shows the weak probe field. The detunings of the probe and couple fields to the respective resonance frequencies are denoted by Δ_p and Δ_c, respectively

$$\partial_t \widehat{\rho} = \frac{1}{i\hbar}\left[\widehat{H}, \widehat{\rho}\right] + \widehat{\rho}_s \,,$$
$$\frac{1}{c}\partial_t \Omega_p + \partial_y \Omega_p = i\eta a_{31}\rho_{31} \,. \tag{5.5}$$

The interaction Hamiltonian is given by

$$\widehat{H} = -\frac{\hbar}{2}\begin{pmatrix} 0 & 0 & a_{31}\Omega_p^* \\ 0 & -2(\Delta_p - \Delta_c) & a_{32}\Omega_c^* \\ a_{31}\Omega_p & a_{32}\Omega_c & -2\Delta_p \end{pmatrix}, \tag{5.6}$$

and the decoherence matrix

$$\widehat{\rho_s} = -\begin{pmatrix} -a_{31}^2\Gamma\rho_{33} & \gamma_{21}\rho_{12} & \left(\gamma_{31} + \frac{\Gamma}{2}\right)\rho_{13} \\ \gamma_{21}\rho_{21} & -a_{32}^2\Gamma\rho_{33} & \left(\gamma_{32} + \frac{\Gamma}{2}\right)\rho_{23} \\ \left(\gamma_{31} + \frac{\Gamma}{2}\right)\rho_{31} & \left(\gamma_{32} + \frac{\Gamma}{2}\right)\rho_{32} & \Gamma\rho_{33} \end{pmatrix}. \tag{5.7}$$

The initial and boundary conditions are

$$\rho_{ij}(0, y) = \delta_{i1}\delta_{1j} \,,$$
$$\Omega_p(0, y) = 0 \,,$$
$$\Omega_p(t, 0) = \varphi Exp\left[-\left(\frac{t}{T}\right)^2\right]. \tag{5.8}$$

In the equations above $\rho_{jk} = A_j A_k^*$ for $\{j, k\} \in \{1, 2, 3\}$ are the density matrix elements of $\widehat{\rho}$ for the nuclear wave function $|\psi\rangle = A_1|\frac{5}{2}, \frac{5}{2}\rangle + A_2|\frac{5}{2}, \frac{3}{2}\rangle + A_3|\frac{3}{2}, \frac{3}{2}\rangle$ with the nuclear hyperfine levels shown in Fig. 5.2b. Furthermore, $(a_{31}, a_{32}) = (\sqrt{2/3}, -2/\sqrt{15})$ are the corresponding Clebsch-Gordan coefficients. $\widehat{\rho_s}$ describes the decoherence due to spin relaxation $(\gamma_{31}, \gamma_{32}, \gamma_{21}) = 2\pi \times (251, 108, 30)$ Hz [32] and spontaneous decay $\Gamma = 0.1$ mHz [5].[7] The parameter η is defined as $\eta = \frac{\Gamma\xi}{2L}$, where ξ represents the effective resonant thickness and $L = 10$ mm the thickness of the target, respectively. Further notations are $\Omega_{p(c)}$ for the Rabi frequency of the probe (couple) VUV radiation [2, 39, 40], $\Delta_{p(c)}$ for the laser detuning of probe (couple) field and c the speed of light.

Since the electric quadrupole splitting of the ground states can be experimentally determined via standard nuclear magnetic resonance (NMR) techniques [32], one can set the two detunings of the coupling and probe fields to be identical but unknown initially, i.e., $\Delta_c = \Delta_p = \Delta$, also known as two photon resonance condition[8]. We

[7] A recent experiment [4] showed the half-lifetime of ^{229m}Th is 6 ± 1 h which corresponds to $\Gamma = 0.032$ mHz.

[8] A legitimate question may arise [41] should Ω_p and Ω_c have any phase correlation? The quick answer might be 'no'. Regardless of whether using VUV laser or synchrotron radiation as probe, the time spectra should be the same. This is because what we need is just to modify the dispersion relation of Ω_p, and this is already changed by Ω_c before the arrival of Ω_p. However, in practice the

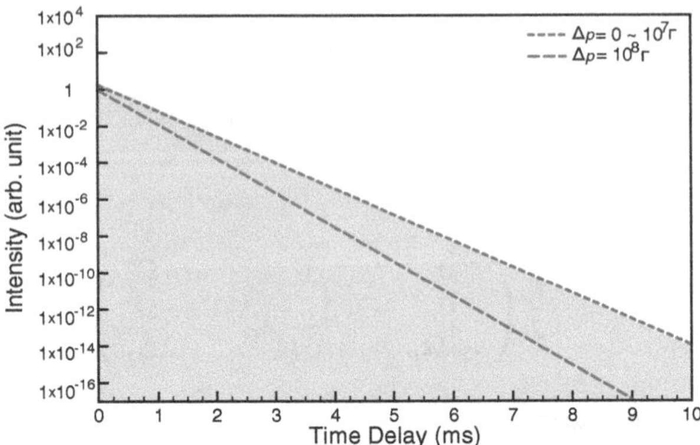

Fig. 5.4 The time spectra of the nuclear forward scattered signal without couple field. The detunings take values between $0 \leq \Delta_p \leq 10^{10}\Gamma$. The *yellow* filled area below the *red short-dashed line* delimits the region $0 \leq \Delta_p \leq 10^7\Gamma$. The values for $\Delta_p = 10^9\Gamma$ and $\Delta_p = 10^{10}\Gamma$ are smaller than the displayed scale

numerically solve Eq. (5.5) with $\xi = 10^6$, Ω_c with a laser intensity[9] of 2 kW/cm^2 [40]. An incident $\Omega_p(t, 0)$ pulse duration $T = 10\,\mu$s and $\varphi \ll \Gamma < \Omega_c$ are used to prevent any Rabi oscillation such that the dynamics of the probe pulse evolves in the perturbation region.[10]

We scan the region $0 \leq \Delta_p \leq 10^{10}\Gamma$ and present the time spectra for the setup in Fig. 5.3a in Fig. 5.4. In the absence of the couple field, the probe only interacts with a two-level nuclear system, and the corresponding time spectra are less sensitive to the detuning Δ_p. For $\Delta_p = 10^9\Gamma$ and $\Delta_p = 10^{10}\Gamma$, the probe field will not create any nuclear excitation and just passes through the crystal. The ^{229}Th nuclei will start to coherently scatter probe photons when $\Delta_p \leq 10^8\Gamma$, and with smaller detunings the time spectra become identical for $\Delta_p \leq 10^7\Gamma$. Essentially we could approach the wanted energy of ^{229m}Th with the precision of $10^7\Gamma$, i.e. about 1 kHz, by measuring the signal photons scattered in the forward direction. As a disadvantage, since the spectra do not present any specific feature to confirm the excitation of the isomer, background from other unwanted electronic processes may present an identical scattering pattern at probe laser frequencies far away from the nuclear resonance.

We turn now to the two-field setup presented in Fig. 5.3b, which is more detuning-sensitive and provides specific identification features in the NFS time spectra. In Fig. 5.5 we present our results for the electromagnetically modified forward scat-

(Footnote 8 continued)
two photon resonance condition indicates there should be a phase correlation between two beams. One may further argue NFS is a broadband excitation which means without phase lock a fluctuant Δ_c still can hit some Δ_p within the bandwidth of Ω_p. (Just like using a needle to fiddle with a comb back and forth.) The ^{229}Th:Crystal may provide a good platform to test this.

9 A higher laser intensity might be needed in view of the recent experimental results in Ref. [4].

10 In the perturbation region, the pattern of NFS time spectra will be φ-independent.

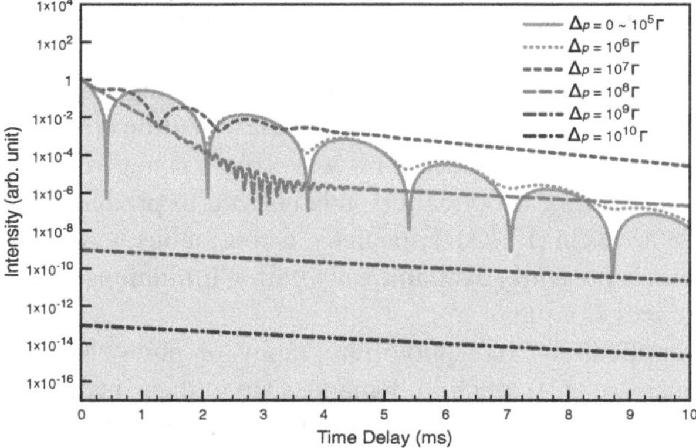

Fig. 5.5 The time spectra of the electromagnetically modified nuclear forward scattered signal with couple field. The detunings ($\Delta_c = \Delta_p$) take values between $0 \leq \Delta_p \leq 10^{10}\Gamma$. The *yellow filled area* below the *orange solid line* delimits the region $0 \leq \Delta_p \leq 10^5\Gamma$

tering spectra. With the influence of Ω_c the spectra are much more sensitive to the detuning, and probe photons start to interact with the ^{229}Th nuclei already at $\Delta \leq 10^{10}\Gamma$. The enhanced sensitivity of the electromagnetically modified setup in Fig. 5.3b is due to the detuning-sensitive dispersion relation which leads to a unique time spectrum for each combination of incident probe field and couple field strength and detuning. A comparison between Figs. 5.4 and 5.5 shows that the couple field has introduced additional beatings in the spectrum which can serve as a clear signature of the nuclear excitation. This electromagnetically induced quantum beat can be understood as following. The purpose of applying the couple field is to create the so-called dressed states [39], i.e., two linear superpositions of the $|\frac{5}{2}, \frac{3}{2}\rangle$ and $|\frac{3}{2}, \frac{3}{2}\rangle$ nuclear states. This splits the nuclear resonance driven by the probe field into a doublet via the Autler-Townes effect [38], giving rise to electromagnetically modified time spectra, and the quantum beat that can be used as signature of the nuclear excitation. Furthermore, a significant advantage of our scheme is that the shapes of the spectra are not identical until $\Delta \leq 10^5\Gamma$. A fit of the theoretical and experimental curves can therefore be employed to determine the isomeric transition frequency. By scanning the detunings Δ, i.e., by varying the known frequencies of probe and couple simultaneously, several Δ-dependent NFS time spectra can be measured. A fit with the theoretical curves in Fig. 5.5 allows the determination of the detuning value and thus of the ^{229g}Th$\leftrightarrow$^{229m}Th transition frequency down to a precision of $10^5\Gamma$, i.e., 10 Hz. Here we clearly see the two-field setup provides the opportunity to significantly narrow the uncertainty of the measured isomeric energy. We further emphasize the advantage of this method is that one will roughly know the isomeric energy even if only one time spectrum with some particular Δ is measured, i.e., no complete Δ scan is needed. This is very different from the common quantum metrology, e.g., Ramsey schemes where one has to compare relative quantum responses of different detunings.

5.3.2.2 State-of-the-Art VUV Laser Around 160 nm

Modern laser technology provides photon sources covering wavelengths from sub-200 nm to tens of μm. The wanted thorium isomeric transition is unfortunately located at a very special edge around 160 nm. This wavelength is not well covered due to the lack of transparent nonlinear optical crystals utilized to produce VUV photons via high harmonic generation (HHG). Especially, a continuous wave (CW) laser source in the VUV region is presently available only within limitations. Here we would like to discuss this practical issue.

Because of applications like photolithography or photoelectron spectroscopy, VUV light sources are very much in demand. Due to these needs, the $KBe_2BO_3F_2$ (KBBF) crystals [42, 43] have been successfully developed for generating narrow-band VUV radiation via harmonic generation. This crystal provides wide transparency and large birefringence necessary for phase matched frequency conversion processes in this frequency region [44, 45]. In particular, a quasi-CW laser based on KBBF crystal with a 10 MHz repetition rate and 20 ps pulse duration has been achieved [45]. A CW coupling VUV laser at around 160 nm wavelength could also be generated by sum frequency mixing in metal vapors or driving a KBBF crystal with Ti:Sapphire laser [44, 46]. The circularly polarized probe laser on the other hand may be generated via nonlinear sum-frequency mixing [47], or a harmonic of a VUV frequency comb [48, 49] around the isomeric wavelength. Another option is offered by free electron lasers (FEL). FELs can essentially generate photons covering a wide frequency spectrum, and the VUV region is clearly available. Recently, a notable table-top laser-driven soft-X-ray undulator source was built [50]. This table-top design with some proper parameters is in principle also able to deliver VUV photon beams.

In the absence of a CW coupling field for the electromagnetically modified forward scattering scheme, one can use a moderate external magnetic field (B = 1 G) to split two degenerate $\Delta m = 0$ transitions, for instance $|\frac{5}{2}, \frac{1}{2}\rangle \leftrightarrow |\frac{3}{2}, \frac{1}{2}\rangle$ and $|\frac{5}{2}, -\frac{1}{2}\rangle \leftrightarrow |\frac{3}{2}, -\frac{1}{2}\rangle$ in Fig. 5.2a, into a doublet due to the different hyperfine energy shifts introduced by the magnetic dipole interaction. Numerical results for ^{229}Th:CaF_2 and ^{229}Th:$LiCaAlF_6$ crystals for which the calculation of the magnetic hyperfine splitting are straightforward show that the detuning sensitivity goes as far as $\Delta_p = 10^7\Gamma$, i.e., one could determine the nuclear transition frequency within 1 kHz with the Zeeman shift method. In the following we investigate this scheme in more detail.

5.3.2.3 NFS with Two ^{229}Th:CaF_2 Crystals Under the Action of a Magnetic Field

Owing to the present difficult access to CW VUV light sources, we demonstrate other ways to measure the energy of ^{229m}Th level via nuclear forward scattering. As showed in Fig. 5.6 we consider a two-crystals system [51–53] and apply an external

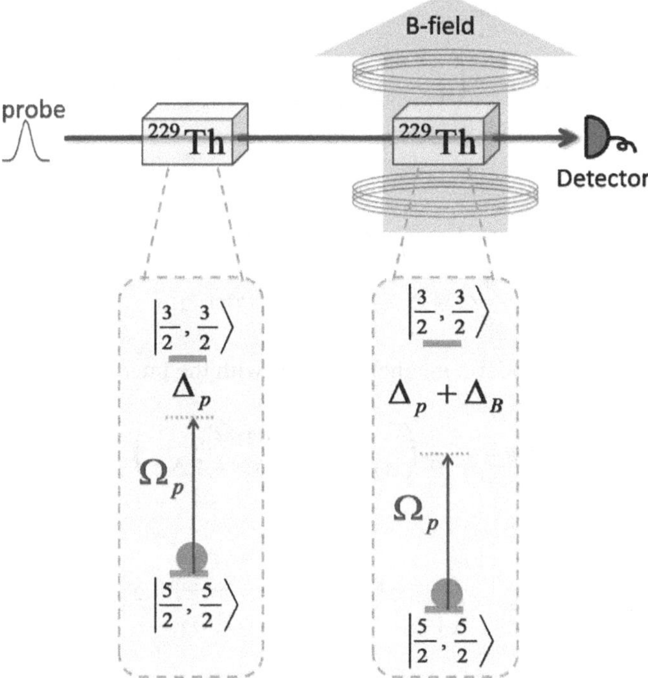

Fig. 5.6 A two ^{229}Th:CaF$_2$ target setup. A left circularly polarized probe field drives the $|\frac{5}{2},\frac{5}{2}\rangle \leftrightarrow |\frac{3}{2},\frac{3}{2}\rangle$ isomeric transitions in both crystals. The detuning of the probe to the unperturbed resonance frequency is denoted by Δ_p. Δ_B denotes the total Zeeman shift due to the external magnetic field B

magnetic field to one of the crystals. All nuclei are assumed to equally populate $|\frac{5}{2},-\frac{5}{2}\rangle$ and $|\frac{5}{2},\frac{5}{2}\rangle$ ground states due to cooling. A left circularly polarized VUV probe pulse sequentially impinges on the two crystals and drives the two originally identical $\Delta m = -1$ transitions $|\frac{5}{2},\frac{5}{2}\rangle \leftrightarrow |\frac{3}{2},\frac{3}{2}\rangle$. The two transitions will be split due to the Zeeman effect that introduces the shift Δ_B, therefore one could anticipate the existence of quantum beat in the NFS time spectra.

We study the two-crystals system in Fig. 5.6 by numerically and sequentially solving Maxwell-Bloch equations [39] for the density matrices $\widehat{\alpha}(t)$ and $\widehat{\beta}(t)$:

$$\partial_t\widehat{\alpha} = \frac{1}{i\hbar}\left[\widehat{H}_A, \widehat{\alpha}\right] + \widehat{\alpha}_s\,,$$
$$\frac{1}{c}\partial_t\Omega_{p1} + \partial_y\Omega_{p1} = i\eta a_{21}\alpha_{21}\,, \tag{5.9}$$

for the first crystal with the B-field-free interaction Hamiltonian

$$\widehat{H}_A = -\frac{\hbar}{2}\begin{pmatrix} 0 & a_{21}\Omega^*_{p1} \\ a_{21}\Omega_{p1} & -2\Delta_p \end{pmatrix}, \tag{5.10}$$

and the decoherence matrix

$$\widehat{\alpha}_s = -\begin{pmatrix} -\Gamma\alpha_{22} & \left(\gamma_{21}+\frac{\Gamma}{2}\right)\alpha_{12} \\ \left(\gamma_{21}+\frac{\Gamma}{2}\right)\alpha_{21} & \Gamma\alpha_{22} \end{pmatrix}, \tag{5.11}$$

and

$$\begin{aligned} \partial_t\widehat{\beta} &= \frac{1}{i\hbar}\left[\widehat{H}_B, \widehat{\beta}\right] + \widehat{\beta}_s\,, \\ \frac{1}{c}\partial_t\Omega_{p2} + \partial_y\Omega_{p2} &= i\eta a_{21}\beta_{21}\,, \end{aligned} \tag{5.12}$$

for the second crystal under a magnetic field **B** with the interaction Hamiltonian

$$\widehat{H}_B = -\frac{\hbar}{2}\begin{pmatrix} 0 & a_{21}\Omega^*_{p2} \\ a_{21}\Omega_{p2} & -2\Delta_p - 2\Delta_B \end{pmatrix}, \tag{5.13}$$

and the decoherence matrix

$$\widehat{\beta}_s = -\begin{pmatrix} -\Gamma\beta_{22} & \left(\gamma_{21}+\frac{\Gamma}{2}\right)\beta_{12} \\ \left(\gamma_{21}+\frac{\Gamma}{2}\right)\beta_{21} & \Gamma\beta_{22} \end{pmatrix}. \tag{5.14}$$

The initial and boundary conditions are

$$\begin{aligned} \alpha_{ij}(0, y) &= \delta_{i1}\delta_{1j}\,, \\ \Omega_{p1}(0, y) &= 0\,, \\ \Omega_{p1}(t, 0) &= \varphi Exp\left[-\left(\frac{t}{T}\right)^2\right], \\ \beta_{ij}(0, y) &= \delta_{i1}\delta_{1j}\,, \\ \Omega_{p2}(0, y) &= 0\,, \\ \Omega_{p2}(t, 0) &= \Omega_{p1}(t, L)\,. \end{aligned} \tag{5.15}$$

All parameters are the same as the ones used in Eq. (5.5). In the equations above $\alpha_{jk} = A_j A_k^*$ for $\{j, k\} \in \{1, 2\}$ are the density matrix elements of $\widehat{\alpha}$ for the nuclear wave function $|\psi\rangle = A_1|\frac{5}{2}, \frac{5}{2}\rangle + A_2|\frac{3}{2}, \frac{3}{2}\rangle$, and $\beta_{jk} = B_j B_k^*$ for $\{j, k\} \in \{1, 2\}$ are the density matrix elements of $\widehat{\beta}$ for the nuclear wave function $|\psi\rangle = B_1|\frac{5}{2}, \frac{5}{2}\rangle + B_2|\frac{3}{2}, \frac{3}{2}\rangle$ with the nuclear hyperfine levels shown in Fig. 5.2b. The Zeeman shift $\Delta_B = \left(\frac{5}{2}\mu_g + \frac{3}{2}\mu_m\right)\mathrm{B}/\hbar$ and $(\mu_g, \mu_m, \mathrm{B}) = (0.45\mu_N, 0.08\mu_N, 10^{-4}\mathrm{T})$ are used, where $\mu_N = 5.05\times 10^{-27}$ (J/T) is the nuclear magneton [5]. We make the assumption that the population of $|\frac{3}{2}, \frac{3}{2}\rangle$ state only decays to $|\frac{5}{2}, \frac{5}{2}\rangle$ due to the large Clebsch-Gordan coefficient $a_{21}^2 = \frac{2}{3}$ and the superradiant speed-up effect. Thus, only a two-level system is involved for each crystal. In order to obtain the total scattered

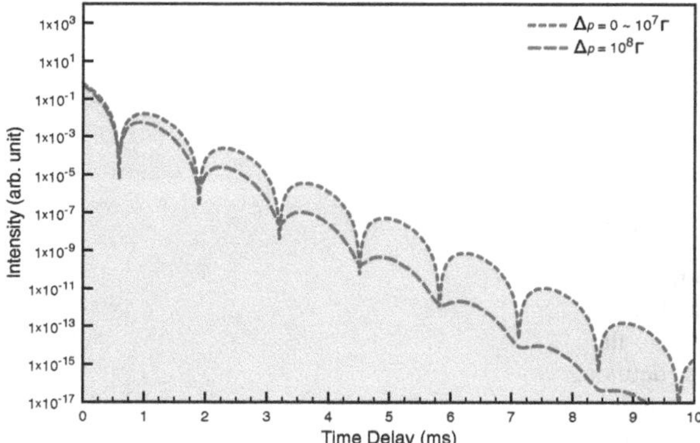

Fig. 5.7 The time spectra of the nuclear forward scattered signal with two cooled ^{229}Th:CaF$_2$ crystals. One of the crystals is under a static magnetic field of 1 Gauss. The detunings take values between $0 \leq \Delta_p \leq 10^{10}\Gamma$. The *yellow filled area* below the *red short-dashed line* delimits the region $0 \leq \Delta_p \leq 10^7\Gamma$. The values for $\Delta_p = 10^9\Gamma$ and $\Delta_p = 10^{10}\Gamma$ are smaller than the displayed scale

field intensity, we sequentially solve Eqs. (5.9) and (5.12) for both crystals using the scattered field of first crystal as the incoming field for the second one under a magnetic field of 1 Gauss.

The results are showed in Fig. 5.7. The tendency is similar to Fig. 5.4 in which nuclei start to scatter photons at $\Delta_p = 10^8\Gamma$, and the time spectra converge on $\Delta_p = 10^7\Gamma$. One can identify the ^{229m}Th energy with a very clear signature given by the quantum beat induced by a very weak magnetic field of 1 Gauss. A very small 1 kHz uncertainty can also be reached by this method with a 10 μs incident probe pulse for the first ^{229}Th:CaF$_2$ target.

5.3.2.4 NFS with One ^{229}Th:LiCaAlF$_6$ Crystal Under the Action of a Magnetic Field

Here, we present another magnetic scheme to measure the isomeric transition energy. In Fig. 5.8 a static magnetic field is applied to a ^{229}Th:LiCaAlF$_6$ target. All nuclei are assumed to be equally populating the two ground states $|\frac{5}{2}, -\frac{1}{2}\rangle$ and $|\frac{5}{2}, \frac{1}{2}\rangle$ due to cooling. A linearly polarized VUV probe pulse impinges on the target and drives two $\Delta m = 0$ transitions $|\frac{5}{2}, \frac{1}{2}\rangle \leftrightarrow |\frac{3}{2}, \frac{1}{2}\rangle$ and $|\frac{5}{2}, -\frac{1}{2}\rangle \leftrightarrow |\frac{3}{2}, -\frac{1}{2}\rangle$. The original single absorption line in the energy domain will also be split by the Zeeman effect giving raise to a quantum beat in the NFS time spectra.

We also neglect the decays to other ground states and consider only four levels in this system. The Maxwell-Bloch equations [39] read:

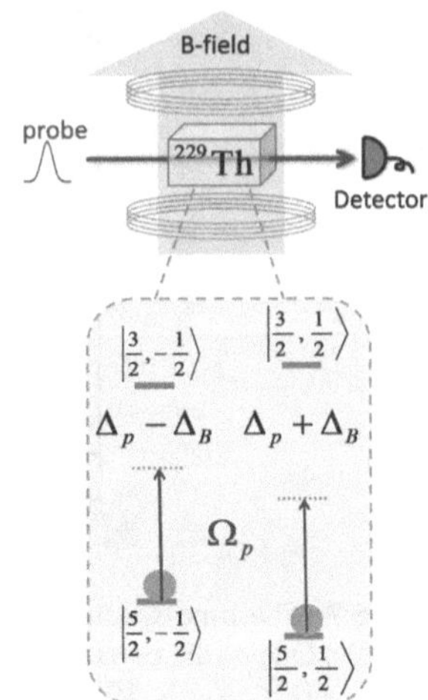

Fig. 5.8 NFS setup in one ^{229}Th:LiCaAlF$_6$ crystal with an external magnetic field **B**. A linearly polarized VUV probe pulse impinges on the crystal and drives two $\Delta m = 0$ transitions $|\frac{5}{2},\frac{1}{2}\rangle \leftrightarrow |\frac{3}{2},\frac{1}{2}\rangle$ and $|\frac{5}{2},-\frac{1}{2}\rangle \leftrightarrow |\frac{3}{2},-\frac{1}{2}\rangle$. The blue arrow shows the weak probe field, and Δ_B is the Zeeman shift due to the magnetic field **B**. The detuning of the probe field to the original unshifted resonance frequency is denoted by Δ_p

$$\partial_t\widehat{\rho} = \frac{1}{i\hbar}\left[\widehat{H},\widehat{\rho}\right] + \widehat{\rho}_s\,,$$
$$\frac{1}{c}\partial_t\Omega_p + \partial_y\Omega_p = i\eta\left(a_{31}\rho_{31} + a_{42}\rho_{42}\right)\,, \tag{5.16}$$

with the interaction Hamiltonian

$$\widehat{H} = -\frac{\hbar}{2}\begin{pmatrix} 0 & 0 & a_{31}\Omega_p^* & 0 \\ 0 & 0 & 0 & a_{42}\Omega_p^* \\ a_{31}\Omega_p & 0 & -2\left(\Delta_p-\Delta_B\right) & 0 \\ 0 & a_{42}\Omega_p & 0 & -2\left(\Delta_p+\Delta_B\right) \end{pmatrix}, \tag{5.17}$$

and the decoherence matrix

$$\widehat{\rho}_s = -\begin{pmatrix} -\Gamma\left(a_{31}^2\rho_{33}+a_{41}^2\rho_{44}\right) & 0 & \left(\gamma_{31}+\frac{\Gamma}{2}\right)\rho_{13} & \frac{\Gamma}{2}\rho_{14} \\ 0 & -\Gamma\left(a_{32}^2\rho_{33}+a_{42}^2\rho_{44}\right) & \frac{\Gamma}{2}\rho_{23} & \left(\gamma_{42}+\frac{\Gamma}{2}\right)\rho_{24} \\ \left(\gamma_{31}+\frac{\Gamma}{2}\right)\rho_{31} & \frac{\Gamma}{2}\rho_{32} & \Gamma\rho_{33} & \Gamma\rho_{34} \\ \frac{\Gamma}{2}\rho_{41} & \left(\gamma_{42}+\frac{\Gamma}{2}\right)\rho_{42} & \Gamma\rho_{43} & \Gamma\rho_{44} \end{pmatrix}, \tag{5.18}$$

where the decoherence rates[11] are $\gamma_{31} = \gamma_{42} = 2\pi\times 150\,\text{Hz}$ [32], the Zeeman shift $\Delta_B = \left(\frac{1}{2}\mu_g + \frac{1}{2}\mu_m\right)\mathrm{B}/\hbar$ and $(a_{31}, a_{32}, a_{41}, a_{42}) = (-\sqrt{2/5}, \sqrt{1/5}, \sqrt{1/5}, -\sqrt{2/5})$ are the corresponding Clebsch-Gordan coefficients. In the equations above $\rho_{jk} = A_j A_k^*$ for $\{j, k\} \in \{1, 2, 3, 4\}$ are the density matrix elements of $\widehat{\rho}$ for the nuclear wave function $|\psi\rangle = A_1|\frac{5}{2},-\frac{1}{2}\rangle + A_2|\frac{5}{2},\frac{1}{2}\rangle + A_3|\frac{3}{2},-\frac{1}{2}\rangle + A_4|\frac{3}{2},\frac{1}{2}\rangle$ with the nuclear hyperfine levels shown in Fig. 5.2a.

[11] We adopt the spin-spin relaxation rate of ^{229}Th:CaF$_2$ [32] assuming the decoherence rate to be on the same order of magnitude for the two types of crystals. In fact we merely consider γ_{31} and γ_{42} because only the coherences ρ_{31} and ρ_{42} will be built up by the Ω_p pulse.

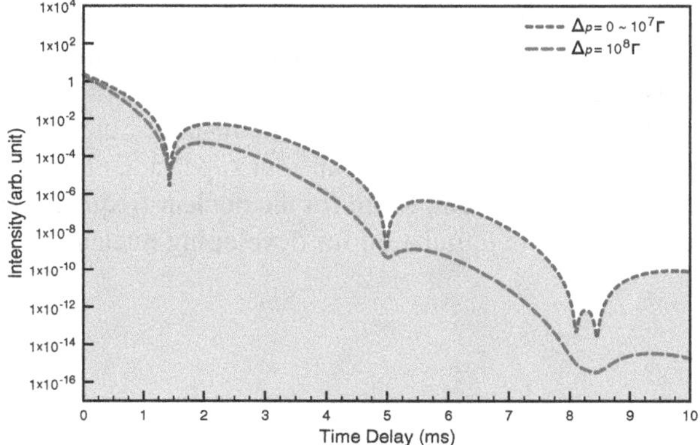

Fig. 5.9 The time spectra of the nuclear forward scattered signal by applying an external magnetic field to a cooled ^{229}Th:LiCaAlF$_6$ target. The detunings take values between $0 \leq \Delta_p \leq 10^{10}\Gamma$. The *yellow filled area* below the *red short-dashed line* delimits the region $0 \leq \Delta_p \leq 10^7\Gamma$. The values for $\Delta_p = 10^9\Gamma$ and $\Delta_p = 10^{10}\Gamma$ are smaller than the displayed scale

The results are presented in Fig. 5.9. The quantum beat in this case is less frequent due to the smaller spin projection of ground and isomeric states. Even so, one can still measure the thorium isomeric transition energy with a small uncertainty of 1 kHz in a ^{229}Th:LiCaAlF$_6$ crystal.

5.4 Summary

In summary, the NFS-based time spectra in the forward direction can be used as an efficient tool to determine the hidden ^{229}Th isomeric transition frequency with small uncertainties of $1\,\text{kHz} \sim 10\,\text{Hz}$ by using a $10\,\mu\text{s}$ probe pulse. Our results show an improvement of many orders of magnitude compared to the traditional indirect schemes [9]. Also, a two-field setup provides both a signature of the nuclear excitation and enhanced sensitivity to the field detuning from the resonance. This scheme demonstrates a novel way to solve the problem of identifying the isomeric transition energy. Additionally, due to the currently difficult access to a CW VUV laser, we suggest the following sequence to measure the ^{229m}Th energy:

1. The typical indirect method, e.g., γ-ray spectroscopy [9].
2. NFS with the magnetically induced quantum beat demonstrated in Sect. 5.3.2.3 and Sect. 5.3.2.4.
3. NFS with the electromagnetically induced quantum beat demonstrated in Sect. 5.3.2.1.

Here, the first step provides an uncertainty of 1 eV, and the second step further narrows the uncertainty down to approximately 1 peV. The last step of our two-field scheme can be used to search the ^{229m}Th energy in a much narrower range such that only a slightly tunable CW VUV laser is needed. Eventually, the isomeric energy will be pinpointed by our two-field scheme with a small uncertainty of only 10 feV. Once the isomeric transition frequency has been identified, the fluorescence clock interrogation scheme using the incoherent response emitted at an angle to the excitation pulse direction (so that no transient superradiant enhancement of the narrow linewidth occurs) can be employed for the nuclear frequency standard. Also, our two-field scheme lays the foundation for developing nuclear quantum optics in ^{229}Th nuclei.

References

1. E. Peik, C. Tamm, Nuclear laser spectroscopy of the 3.5 eV transition in ^{229}Th. Europhys. Lett. **61** (2003)
2. W.-T. Liao, S. Das, C.H. Keitel, A. Pálffy, Coherence enhanced optical determination of the ^{229}Th isomeric transition. Phys. Rev. Lett. **109**, 262502 (2012)
3. S.G. Porsev, V.V. Flambaum, E. Peik, C. Tamm, Excitation of the isomeric ^{229m}Th nuclear state via an electronic bridge process in ^{229}Th^{+}. Phys. Rev. Lett. **105**, 182501 (2010)
4. X. Zhao, Y.N. Martinez de Escobar, R. Rundberg, E.M. Bond, A. Moody, D.J. Vieira, Observation of the deexcitation of the $^{229\mathbf{m}}$Th nuclear isomer. Phys. Rev. Lett. **109**, 160801 (2012)
5. E.V. Tkalya, Proposal for a nuclear gamma-ray laser of optical range. Phys. Rev. Lett. **106**, 162501 (2011)
6. L. Hau, S. Harris, Z. Dutton, C. Behroozi, Light speed reduction to 17 metres per second in an ultracold atomic gas. Nature **397**, 594 (1999)
7. C.W. Reich, R.G. Helmer, Energy separation of the doublet of intrinsic states at the ground state of ^{229}Th. Phys. Rev. Lett. **64**, 271 (1990)
8. K. Gulda, W. Kurcewicz, A. Aas, M. Borge, D. Burke, B. Fogelberg, I. Grant, E. Hagebø, N. Kaffrell, J. Kvasil, G. Løvhøiden, H. Mach, A. Mackova, T. Martinez, G. Nyman, B. Rubio, J. Tain, O. Tengblad, T. Thorsteinsen, The nuclear structure of ^{229}Th. Nucl. Phys. A **703**, 45 (2002)
9. B.R. Beck, J.A. Becker, P. Beiersdorfer, G.V. Brown, K.J. Moody, J.B. Wilhelmy, F.S. Porter, C.A. Kilbourne, R.L. Kelley, Energy splitting of the ground-state doublet in the nucleus ^{229}Th. Phys. Rev. Lett. **98**, 142501 (2007)
10. B.R. Beck, C.Y. Wu, P. Beiersdorfer, G.V. Brown, J.A. Becker, K.J. Moody, J.B. Wilhelmy, F.S. Porter, C.A. Kilbourne, R.L. Kelley, Improved value for the energy splitting of the ground-state doublet in the, nucleus 229mTh. LLNL-PROC-415170 (2009)
11. S.J. Goldstein, M.T. Murrell, R.W. Williams, Half-life of ^{229}Th. Phys. Rev. C **40**, 2793 (1989)
12. L. Kroger, C. Reich, Features of the low-energy level scheme of 229Th as observed in the $\alpha - decay$ of 233u. Nucl. Phys. A **259**, 29 (1976)
13. R.G. Helmer, C.W. Reich, An excited state of ^{229}Th at 3.5 eV. Phys. Rev. C **49**, 1845 (1994)
14. S. Sakharov, On the energy of the 3.5-ev level in 229Th. Phys. At. Nucl. **73**, 1 (2010)
15. D.G. Burke, P.E. Garrett, T. Qu, R.A. Naumann, Additional evidence for the proposed excited state at $\leq$5 eV in ^{229}Th. Phys. Rev. C **42**, R499 (1990)
16. G.M. Irwin, K.H. Kim, Observation of electromagnetic radiation from deexcitation of the ^{229}Th isomer. Phys. Rev. Lett. **79**, 990 (1997)
17. D.S. Richardson, D.M. Benton, D.E. Evans, J.A.R. Griffith, G. Tungate, Ultraviolet photon emission observed in the search for the decay of the ^{229}Th isomer. Phys. Rev. Lett. **80**, 3206–3208 (1998)
18. S.B. Utter, P. Beiersdorfer, A. Barnes, R.W. Lougheed, J.R. Crespo López-Urrutia, J.A. Becker, M.S. Weiss, Reexamination of the optical gamma ray decay in ^{229}tTh. Phys. Rev. Lett. **82**, 505 (1999)

19. R.W. Shaw, J.P. Young, S.P. Cooper, O.F. Webb, Spontaneous ultraviolet emission from ^{233}Uranium/^{229}Thorium samples. Phys. Rev. Lett. **82**, 1109 (1999)
20. F.F. Karpeshin, I.M. Band, M.B. Trzhaskovskaya, A. Pastor, On the question of electron bridge for the 3.5-ev isomer of ^{229}Th. Phys. Rev. Lett. **83**, 1072 (1999)
21. G.M. Irwin, K.H. Kim, Irwin and kim reply: Phys. Rev. Lett. **83**, 1073 (1999)
22. V. Barci, G. Ardisson, G. Barci-Funel, B. Weiss, O. El Samad, R.K. Sheline, Nuclear structure of^{229}Th from γ-ray spectroscopy study of^{233}U α-particle decay. Phys. Rev. C **68**, 034329 (2003)
23. Z.O. Guimarães Filho, O. Helene, Energy of the $3/2^+$ state of ^{229}Th reexamined. Phys. Rev. C **71**, 044303 (2005)
24. C.J. Campbell, A.V. Steele, L.R. Churchill, M.V. DePalatis, D.E. Naylor, D.N. Matsukevich, A. Kuzmich, M.S. Chapman, Multiply charged thorium crystals for nuclear laser spectroscopy. Phys. Rev. Lett. **102**, 233004 (2009)
25. C.J. Campbell, A.G. Radnaev, A. Kuzmich, V.A. Dzuba, V.V. Flambaum, A. Derevianko, Single-ion nuclear clock for metrology at the 19th decimal place. Phys. Rev. Lett. **108**, 120802 (2012)
26. S. Kishimoto, Y. Yoda, M. Seto, Y. Kobayashi, S. Kitao, R. Haruki, T. Kawauchi, K. Fukutani, T. Okano, Observation of nuclear excitation by electron transition in ^{197}Au with synchrotron x rays and an avalanche photodiode. Phys. Rev. Lett. **85**, 1831 (2000)
27. T.T. Inamura, H. Haba, Search for a "3.5-eV isomer" in ^{229}Th in a hollow-cathode electric discharge. Phys. Rev. C **79**, 034313 (2009)
28. L.v.d. Wense, P.G. Thirolf, D. Kalb, M. Laatiaoui, Towards a direct transition energy measurement of the lowest nuclear excitation in ^{229}Th, JINST **8**, P03005 (2013)
29. S. Raeder, V. Sonnenschein, T. Gottwald, I.D. Moore, M. Reponen, S. Rothe, N. Trautmann, K. Wendt, Resonance ionization spectroscopy of thorium isotopes-towards a laser spectroscopic identification of the low-lying 7.6 eV isomer of ^{229}Th. J. Phys. B: At. Mol. Opt. Phys. **44**, 165005 (2011)
30. W.G. Rellergert, S.T. Sullivan, D. DeMille, R.R. Greco, M.P. Hehlen, R.A. Jackson, J.R. Torgerson, E.R. Hudson, Progress towards fabrication of 229 th-doped high energy band-gap crystals for use as a solid-state optical frequency reference. IOP Conf. Ser.: Mater. Sci. Eng. **15**, 012005 (2010)
31. W.G. Rellergert, D. DeMille, R.R. Greco, M.P. Hehlen, J.R. Torgerson, E.R. Hudson, Constraining the evolution of the fundamental constants with a solid-state optical frequency reference based on the ^{229}Th nucleus. Phys. Rev. Lett. **104**, 200802 (2010)
32. G.A. Kazakov, A.N. Litvinov, V.I. Romanenko, L.P. Yatsenko, A.V. Romanenko, M. Schreitl, G. Winkler, T. Schumm, Performance of a 229thorium solid-state nuclear clock. New J. Phys. **14**, 083019 (2012)
33. M.D. Crisp, Propagation of small-area pulses of coherent light through a resonant medium. Phys. Rev. A **1**, 1604 (1970)
34. Y. Shvyd'ko, Nuclear resonant forward scattering of x rays: Time and space picture. Phys. Rev. B **59**, 9132 (1999)
35. Y.V. Shvyd'ko, U. van Bürck, Hybrid forms of beat phenomena in nuclear forward scattering of synchrotron radiation. Hyperfine Interact. **123–124**, 511 (1999)
36. S.L. Ruby, Mössbauer experiments without conventional sources. J. Phys. Colloques **35**, C6–209 (1974)
37. R. Röhlsberger, *Nuclear Condensed Matter Physics with Synchrotron Radiation: Basic Principles, Methodology and Applications* (Springer, Berlin, 2004)
38. S.H. Autler, C.H. Townes, Stark effect in rapidly varying fields. Phys. Rev. **100**, 703 (1955)
39. M.O. Scully, M.S. Zubairy, *Quantum Optics* (Cambridge University Press, Cambridge, 1997)
40. A. Pálffy, J. Evers, C.H. Keitel, Electric-dipole-forbidden nuclear transitions driven by super-intense laser fields. Phys. Rev. C **77**, 044602 (2008)
41. T. Schumm, private communications (2012)
42. C. Chen, Z. Xu, D. Deng, J. Zhang, G.K.L. Wong, B. Wu, N. Ye, D. Tang, The vacuum ultraviolet phase-matching characteristics of nonlinear optical $KBe_2BO_3F_2$ crystal. Appl. Phys. Lett. **68**, 2930 (1996)

43. C. Chen, G. Wang, X. Wang, Z. Xu, Deep-UV nonlinear optical crystal $KBe_2BO_3F_2$–discovery, growth, optical properties and applications. Appl. Phys. B **97**, 9 (2009)
44. T. Togashi, T. Kanai, T. Sekikawa, S. Watanabe, C. Chen, C. Zhang, Z. Xu, J. Wang, Generation of vacuum-ultraviolet light by an optically contacted, prism-coupled $KBe_2BO_3F_2$ crystal. Opt. Lett. **28**, 254 (2003)
45. Y. Nomura, Y. Ito, A. Ozawa, X. Wang, C. Chen, S. Shin, S. Watanabe, Y. Kobayashi, Coherent quasi-cw 153 nm light source at 33 mHz repetition rate. Opt. Lett. **36**, 1758 (2011)
46. J. Nolting, H. Kunze, I. Schütz, R. Wallenstein, Cw coherent vuv radiation generated by resonant sum-frequency mixing in metal vapors. Appl. Phys. B **50**, 331 (1990)
47. R. Irrgang, M. Drescher, F. Gierschner, M. Spieweck, U. Heinzmann, A laser light source for circularly polarized VUV radiation. Meas. Sci. Technol. **9**, 422 (1998)
48. R.J. Jones, K.D. Moll, M.J. Thorpe, J. Ye, Phase-coherent frequency combs in the vacuum ultraviolet via high-harmonic generation inside a femtosecond enhancement cavity. Phys. Rev. Lett. **94**, 193201 (2005)
49. A. Ozawa, J. Rauschenberger, C. Gohle, M. Herrmann, D.R. Walker, V. Pervak, A. Fernandez, R. Graf, A. Apolonski, R. Holzwarth, F. Krausz, T.W. Hänsch, T. Udem, High harmonic frequency combs for high resolution spectroscopy. Phys. Rev. Lett. **100**, 253901 (2008)
50. M. Fuchs, R. Weingartner, A. Popp, Z. Major, S. Becker, J. Osterhoff, I. Cortrie, B. Zeitler, R. Horlein, G.D. Tsakiris, U. Schramm, T.P. Rowlands-Rees, S.M. Hooker, D. Habs, F. Krausz, S. Karsch, F. Gruner, Laser-driven soft-x-ray undulator source. Nat. Phys. **5**, 826 (2009)
51. G.V. Smirnov, U. van Bürck, J. Arthur, S.L. Popov, A.Q.R. Baron, A.I. Chumakov, S.L. Ruby, W. Potzel, G.S. Brown, Nuclear exciton echo produced by ultrasound in forward scattering of synchrotron radiation. Phys. Rev. Lett. **77**, 183 (1996)
52. G.V. Smirnov, U. van Bürck, W. Potzel, P. Schindelmann, S.L. Popov, E. Gerdau, Y.V. Shvyd'ko, H.D. Rüter, O. Leupold, Propagation of nuclear polaritons through a two-target system: Effect of inversion of targets. Phys. Rev. A **71**, 023804 (2005)
53. W.-T. Liao, A. Pálffy, C.H. Keitel, Coherent storage and phase modulation of single hard-x-ray photons using nuclear excitons. Phys. Rev. Lett. **109**, 197403 (2012)

Chapter 6
Conclusions and Outlook

In this thesis, three ideas have been proposed to extend the grasp of quantum optics into nuclear systems and towards higher photon energies to elucidate the possibility of coherently manipulating both nuclei and photons.

In Chap. 3, the feasibility of using STIRAP and two π-pulses to coherently transfer the population between two nuclear ground states in a Λ-type system has been investigated. We considered an accelerated nuclear bunch that interacts with two Lorentz-boosted XFEL pulses, and numerically calculated the NCPT efficiency with an optimization process of XFEL parameters. We found that the parameter regime for which fully coherent x-ray laser pulses can induce population transfer between nuclear levels matches the predicted values for the envisaged XFELO and SXFEL facilities. The challenge for the experimental realization of NCPT and the future of nuclear batteries thus rely on the development of x-ray coherent sources and their conjuncture with ion accelerators, perhaps making use of high-precision table-top solutions for lasers and ion accelerators to be flexibly used at any location around the globe.

In Chap. 4, schemes of coherent storage and phase modulation for single hard x-ray photons have been proposed. By manipulating the hyperfine magnetic field applied to ^{57}Fe nuclei in a NFS setup, the nuclear photon emission can be coherently controlled. Additionally, inspired by the NCPT in Chap. 3, a generation of single hard x-ray photons in a nearly deterministic fashion can be implemented by replacing synchrotron radiation with fully coherent XFEL in a NFS setup under the condition of Eq. (4.25). These x-ray coherent control tools are important milestones for optics and quantum information applications at shorter wavelengths aiming towards more compact future photonic devices.

In Chap. 5, we have theoretically demonstrated that four setups based on the NFS method can be utilized to overcome the present impediment for constructing ^{229}Th nuclear clocks, i.e., the 1 eV uncertainty of the considered clock transition ^{229m}Th$\rightarrow$^{229g}Th line in γ-ray spectroscopy. Following our proposed coherence-enhanced optical determination schemes, one can compare the measured NFS time spectra with theoretical calculations to precisely determine the clock transition

W.-T. Liao, *Coherent Control of Nuclei and X-Rays*, Springer Theses,
DOI: 10.1007/978-3-319-02120-1_6,

energy. It turns out that a two-field scheme reminding of electromagnetically induced transparency (EIT) (see sect. 5.3.2.1) may have a significant impact on identifying the ^{229}Th nuclear clock transitions.

Starting from these three applications, as an outlook many other future directions in nuclear quantum optics emerge, for example:

^{229}Th Nuclear Quantum Optics

Inspired by the extremely narrow linewidth of ^{229m}Th isomer state, a test for the effective pulse shape of Eq. (B.3) can be envisaged. This equation gives the effective laser pulse that may interact with a nucleus as its considered transition linewidth is narrower than the bandwidth of an incident XFEL pulse. Once Eq. (B.3) is experimentally verified, a quasi-CW VUV pulsed laser with a high repetition rate [1] can be used as a CW laser in a ^{229}Th system. Moreover, the implementation of the schemes demonstrated in Chap. 4 with ^{229}Th nuclei may also be promising for future quantum information processes.

Control of Single X-Ray Photons

In addition to quantum information, well controlled single hard x-ray photons may have impact also on other applications and deserve further developments. In particular, due to the large momentum carried by a single hard x-ray photon, a possible combination of nuclear quantum optics and optomechanics may lead to more apparent effects and controllable optics elements for hard x-rays. For instance it may be more promising to implement the proposal of quantum superposition of a mirror [2] with a single hard x-ray photon. Also, quantum interference effects of single hard x-ray photons are also interesting problems to be studied, e.g., the Hong-Ou-Mandel effect [3].

The Interplay between a Nucleus and its Bound Electrons

In an atom, a nucleus interacts with the bound electrons via the electromagnetic force and the overlaps of their quantum wave functions. These effects lead to, e.g., the atomic hyperfine structure [4] and the nuclear decay via so-called internal conversion (IC) [5]. Thus, one could think about controlling the nuclear behavior via manipulating the shell electronic shell wave functions [6], for instance suppressing the IC by putting electrons in states with high orbital angular momentum. Also, the possibility of incoherently exciting nuclei by shining an electron beam on them deserves more attention [7].

Control of Mössbauer Spectroscopy with Nuclear Motions

In a crystal, the resonant nuclear x-ray absorption line is usually not alone but accompanied by two side bands [8]. These two side bands occur due to the creation and annihilation of one phonon in the crystal. One could therefore think about modifying the γ-ray absorption by controlling nuclear motion [9, 10]. This kind of research might lead to new and controllable optics elements for γ-rays.

Synthetic Combinations of Nuclei, Radiation and Some Mesoscopic Systems

Recent experiments [11–14] demonstrated that the interaction between nuclei and radiation can be modified when realized in a cavity. One can think about integrating both an atomic and a nuclear system in a cavity to control one component by manipulating another. Also, in a synthetic quantum information system [15], one can access the data in tiny quantum memories with an single x-ray photon and transfer the data with a single optical photon for a long distance. In such "multi-color" networks, an interface between different colors is paramount [15], and an integrated atom-nucleus system may play the key role.

Nuclear quantum optics—as we have seen there is a vast land of unexplored opportunities which may lead to breakthrough for manipulating atomic nuclei. Also, the applications inspired by these discoveries in the near future are more than promising. The surf is up.

References

1. Y. Nomura, Y. Ito, A. Ozawa, X. Wang, C. Chen, S. Shin, S. Watanabe, Y. Kobayashi, Coherent quasi-cw 153 nm light source at 33 mHz repetition rate. Opt. Lett. **36**, 1758 (2011)
2. W. Marshall, C. Simon, R. Penrose, D. Bouwmeester, Towards quantum superpositions of a mirror. Phys. Rev. Lett. **91**, 130401 (2003)
3. C.K. Hong, Z.Y. Ou, L. Mandel, Measurement of subpicosecond time intervals between two photons by interference. Phys. Rev. Lett. **59**, 2044 (1987)
4. D.J. Griffiths, *Introduction to quantum mechanics* (Pearson Prentice Hall, Upper Saddle River, New Jersy, 2005)
5. J.M. Blatt, V.F. Weisskopf, *Theoretical Nuclear Physics* (Courier Dover Publications, America, 1991)
6. O. Kocharovskaya, R. Kolesov, Y. Rostovtsev, Coherent optical control of Mössbauer spectra. Phys. Rev. Lett. **82**, 3593 (1999)
7. A. Pálffy, J. Evers, C.H. Keitel, Isomer triggering via nuclear excitation by electron capture. Phys. Rev. Lett. **99**, 172502 (2007)
8. M. Seto, Y. Yoda, S. Kikuta, X.W. Zhang, M. Ando, Observation of nuclear resonant scattering accompanied by phonon excitation using synchrotron radiation. Phys. Rev. Lett. **74**, 3828 (1995)
9. Y.V. Radeonychev, M.D. Tokman, A.G. Litvak, O. Kocharovskaya, Acoustically induced transparency in optically dense resonance medium. Phys. Rev. Lett. **96**, 093602 (2006)
10. F. Vagizov, S. Olariu, O. Kocharovskaya, Experimental search for laser-induced effects in 151 Eu and 57 Fe doped crystals. Laser phys. **17**, 734 (2007)
11. R. Röhlsberger, K. Schlage, T. Klein, O. Leupold, Accelerating the spontaneous emission of x rays from atoms in a cavity. Phys. Rev. Lett. **95**, 097601 (2005)
12. R. Röhlsberger, K. Schlage, B. Sahoo, S. Couet, R. Rüffer, Collective lamb shift in single-photon superradiance. Science **328**, 1248 (2010)
13. R. Röhlsberger, H. Wille, K. Schlage, B. Sahoo, Electromagnetically induced transparency with resonant nuclei in a cavity. Nature **482**, 199 (2012)
14. K.P. Heeg, H.C. Wille, K. Schlage, T. Guryeva, I. Uschmann, K.S. Schulze, B. Marx, T.Kämpfer, G.G. Paulus, R.Röhlsberger, J. Evers, Spontaneously generated coherences in the x-ray regime. Phys. Rev. Lett. **111**, 073601 (2013)
15. M.G. Raymer, K. Srinivasan, Manipulating the color and shape of single photons. Phys. Today **65**, 32 (2012)

Appendix A
Supplemental Material for Chapter 2

Steps Between Eqs. (2.31) and (2.32)

$$
\begin{aligned}
f_n(L', M; r) &= \sum_{L=0}^{\infty} R(L; r) \int_0^{2\pi} \int_0^{\pi} \mathbf{Y}_{L'L'}^{M*}(\beta, \phi) \cdot Y_{L0}(\beta) \widehat{\chi}_n \sin \beta d\beta d\phi \\
&= \sum_{L=0}^{\infty} R(L; r) \sum_{m=-L'}^{L'} \sum_{n'=-1}^{1} \widehat{\chi}_{n'}^{*} \cdot \widehat{\chi}_n \mathrm{C}_{L'1}(L', M; m, n') \\
&\quad \times \int_0^{2\pi} \int_0^{\pi} Y_{L'm}^{*}(\beta, \phi) Y_{L0}(\beta) \sin \beta d\beta d\phi \\
&= \sum_{L=0}^{\infty} R(L; r) \sum_{m=-L'}^{L'} \sum_{n'=-1}^{1} \widehat{\chi}_{n'}^{*} \cdot \widehat{\chi}_n \mathrm{C}_{L'1}(L', M; m, n') \delta_{L'L} \delta_{m0} \delta_{n'M} \\
&= R(L'; r) \sum_{n'=-1}^{1} \widehat{\chi}_{n'}^{*} \cdot \widehat{\chi}_n \mathrm{C}_{L'1}(L', M; 0, n') \delta_{n'M} \qquad \text{(A.1)} \\
&= R(L'; r) [\widehat{\chi}_{-1}^{*} \cdot \widehat{\chi}_n \delta_{-1M} \sqrt{\frac{(L' - M)(L' + M + 1)}{2L'(L' + 1)}} \\
&\quad - \widehat{\chi}_{1}^{*} \cdot \widehat{\chi}_n \delta_{1M} \sqrt{\frac{(L' + M)(L' - M + 1)}{2L'(L' + 1)}}] \qquad \text{(A.2)} \\
&= \frac{1}{\sqrt{2}} R(L'; r) \left(\delta_{-1M} \widehat{\chi}_{-1}^{*} \cdot \widehat{\chi}_n - \delta_{1M} \widehat{\chi}_{1}^{*} \cdot \widehat{\chi}_n \right) \qquad \text{(A.3)}
\end{aligned}
$$

W.-T. Liao, *Coherent Control of Nuclei and X-Rays*, Springer Theses,
DOI: 10.1007/978-3-319-02120-1,

Table A.1 Clebsch-Gordan Coefficients $C_{\ell 1}(L, M; m, n)$. We adopt the definition in Ref. [1]

	$n = 1$	$n = 0$	$n = -1$
$L = \ell + 1$	$\sqrt{\frac{(\ell+M)(\ell+M+1)}{(2\ell+1)(2\ell+2)}}$	$\sqrt{\frac{(\ell-M+1)(\ell+M+1)}{(2\ell+1)(\ell+1)}}$	$\sqrt{\frac{(\ell-M)(\ell-M+1)}{(2\ell+1)(2\ell+2)}}$
$L = \ell$	$-\sqrt{\frac{(\ell+M)(\ell-M+1)}{2\ell(\ell+1)}}$	$\frac{M}{\sqrt{\ell(\ell+1)}}$	$\sqrt{\frac{(\ell-M)(\ell+M+1)}{2\ell(\ell+1)}}$
$L = \ell - 1$	$\sqrt{\frac{(\ell-M)(\ell-M+1)}{2\ell(2\ell+1)}}$	$-\sqrt{\frac{(\ell-M)(\ell+M)}{\ell(2\ell+1)}}$	$\sqrt{\frac{(\ell+M+1)(\ell+M)}{2\ell(2\ell+1)}}$

Steps Between Eqs. (2.33) and (2.34)

$$
\begin{aligned}
g_n(L', M; r) &= \sum_{L=0}^{\infty} R(L; r) \int_0^{2\pi} \int_0^{\pi} \mathbf{Y}_{L'L'+1}^{M*}(\beta, \phi) \cdot Y_{L0}(\beta) \widehat{\chi}_n \sin\beta d\beta d\phi \\
&= \sum_{L=0}^{\infty} R(L; r) \sum_{m=-L'-1}^{L'+1} \sum_{n'=-1}^{1} \widehat{\chi}_{n'}^* \cdot \widehat{\chi}_n \mathrm{C}_{L'+11}(L', M; m, n') \\
&\quad \times \int_0^{2\pi} \int_0^{\pi} Y_{L'+1m}^*(\beta, \phi) Y_{L0}(\beta) \sin\beta d\beta d\phi \\
&= \sum_{L=0}^{\infty} R(L; r) \sum_{m=-L'-1}^{L'+1} \sum_{n'=-1}^{1} \widehat{\chi}_{n'}^* \cdot \widehat{\chi}_n \mathrm{C}_{L'+11}(L', M; m, n') \delta_{L'+1L} \delta_{m0} \delta_{n'M} \\
&= R(L'+1; r) \sum_{n'=-1}^{1} \widehat{\chi}_{n'}^* \cdot \widehat{\chi}_n \mathrm{C}_{L'+11}(L', M; 0, n') \delta_{n'M} \qquad \text{(A.4)} \\
&= R(L'+1; r) \left[\delta_{-1M} \widehat{\chi}_{-1}^* \cdot \widehat{\chi}_n \sqrt{\frac{L'}{2(2L'+3)}} + \delta_{1M} \widehat{\chi}_1^* \cdot \widehat{\chi}_n \sqrt{\frac{L'}{2(2L'+3)}} \right]. \qquad \text{(A.5)}
\end{aligned}
$$

Steps Between Eqs. (2.35) and (2.36)

$$
\begin{aligned}
h_n(L', M; r) &= \sum_{L=0}^{\infty} R(L; r) \int_0^{2\pi} \int_0^{\pi} \mathbf{Y}_{L'L'-1}^{M*}(\beta, \phi) \cdot Y_{L0}(\beta) \widehat{\chi}_n \sin\beta d\beta d\phi \\
&= \sum_{L=0}^{\infty} R(L; r) \sum_{m=-L'+1}^{L'-1} \sum_{n'=-1}^{1} \widehat{\chi}_{n'}^* \cdot \widehat{\chi}_n \mathrm{C}_{L'-11}(L', M; m, n') \\
&\quad \times \int_0^{2\pi} \int_0^{\pi} Y_{L'-1m}^*(\beta, \phi) Y_{L0}(\beta) \sin\beta d\beta d\phi \\
&= \sum_{L=0}^{\infty} R(L; r) \sum_{m=-L'+1}^{L'-1} \sum_{n'=-1}^{1} \widehat{\chi}_{n'}^* \cdot \widehat{\chi}_n \mathrm{C}_{L'-11}(L', M; m, n') \delta_{L'-1L} \delta_{m0} \delta_{n'M}
\end{aligned}
$$

$$= R(L'-1;r)\sum_{n'=-1}^{1}\widehat{\chi}_{n'}^{*}\cdot\widehat{\chi}_{n}C_{L'-11}(L',M;0,n')\delta_{n'M} \tag{A.6}$$

$$= R(L'-1;r)\left[\delta_{-1M}\widehat{\chi}_{-1}^{*}\cdot\widehat{\chi}_{n}\sqrt{\frac{L'+1}{2(2L'-1)}}+\delta_{1M}\widehat{\chi}_{1}^{*}\cdot\widehat{\chi}_{n}\sqrt{\frac{L'+1}{2(2L'-1)}}\right] \tag{A.7}$$

Steps Between Eqs. (2.39) and (2.46)

For convenience, we replace L' with L in the following. By substituting Eq. (2.39) into Eq. (2.16), we obtain

$$\begin{aligned}\langle e|\widehat{H}_I(t)|g\rangle &= -\frac{i}{\omega_k}e^{i\Delta_k t}\mathbf{E}_k\cdot\int_V \mathbf{j}^*(\mathbf{r})e^{ikr\cos\beta}d^3\mathbf{r}+\frac{i}{\omega_k}e^{-i\Delta_k t}\mathbf{E}_k^*\cdot\int_V \mathbf{j}(\mathbf{r})e^{-ikr\cos\beta}d^3\mathbf{r}\\ &= -i\frac{\sqrt{2\pi}}{\omega_k}e^{-i\Delta_k t}\int_V d^3\mathbf{r}\,\mathbf{j}^*(\mathbf{r})\cdot\sum_{n=-1,1}(\mathbf{E}_k\cdot\widehat{\chi}_n^*)\\ &\quad\times\left(\widehat{\chi}_{-1}^*\cdot\widehat{\chi}_n\,,\ \widehat{\chi}_1^*\cdot\widehat{\chi}_n\right)\begin{pmatrix}\left(1-\frac{1}{k}\nabla\times\right)\sum_L(i)^L\sqrt{2L+1}j_L(kr)\mathbf{Y}_{LL}^{-1}\\ \left(-1-\frac{1}{k}\nabla\times\right)\sum_L(i)^L\sqrt{2L+1}j_L(kr)\mathbf{Y}_{LL}^{1}\end{pmatrix}.\\ &\quad+H.c.\end{aligned} \tag{A.8}$$

With the following relations [1]

$$\mathbf{Y}_{LL}^{M}(\beta,\phi)=\frac{-i(\mathbf{r}\times\nabla)Y_{LM}(\beta,\phi)}{\sqrt{L(L+1)}}, \tag{A.9}$$

$$j_L(kr)\cong\frac{(kr)^L}{(2L+1)!!},\ \text{for}\ \frac{1}{k}\gg r, \tag{A.10}$$

where Eq. (A.10) is based on that the size of nucleus is smaller than wave lengths of interacting radiation fields, and the symbol !! in Eq. (A.10) denotes the double factorial. For further simplification, we define

$$A_{Mag}^{*}=\int_V r^L(\mathbf{r}\times\nabla Y_{LM})\cdot\mathbf{j}^*(\mathbf{r})d^3\mathbf{r} \tag{A.11}$$

$$=\int_V \mathbf{j}^*(\mathbf{r})\cdot\left[\mathbf{r}\times\left(r^L\nabla Y_{LM}+Y_{LM}\nabla r^L\right)\right]d^3\mathbf{r} \tag{A.12}$$

$$=\int_V \mathbf{j}^*(\mathbf{r})\cdot\left[\mathbf{r}\times\nabla\left(r^L Y_{LM}\right)\right]d^3\mathbf{r} \tag{A.13}$$

$$=-\int_V\left[\mathbf{r}\times\mathbf{j}^*(\mathbf{r})\right]\cdot\nabla\left(r^L Y_{LM}\right)d^3\mathbf{r}. \tag{A.14}$$

In Eq. (A.12), we use $\mathbf{r}\times\nabla r^L=0$. On the other hand,

$$A^*_{Ele} = \int_V \nabla \times \left[r^L (\mathbf{r} \times \nabla Y_{LM})\right] \cdot \mathbf{j}^*(\mathbf{r}) d^3\mathbf{r} \tag{A.15}$$

$$= \int_V \left\{\nabla \times \left[\mathbf{r} \times \nabla \left(r^L Y_{LM}\right)\right]\right\} \cdot \mathbf{j}^*(\mathbf{r}) d^3\mathbf{r} \tag{A.16}$$

$$= \int_V \nabla \cdot \left\{\left[\mathbf{r} \times \nabla \left(r^L Y_{LM}\right)\right] \times \mathbf{j}^*(\mathbf{r})\right\} d^3\mathbf{r} - \int_V \left[\mathbf{r} \times \nabla \left(r^L Y_{LM}\right)\right] \cdot \left[\nabla \times \mathbf{j}^*(\mathbf{r})\right] d^3\mathbf{r} \tag{A.17}$$

$$= \oint_S \left[\mathbf{r} \times \nabla \left(r^L Y_{LM}\right)\right] \times \mathbf{j}^*(\mathbf{r}) d^2\mathbf{r} - \int_V \left[\nabla \times \mathbf{j}^*(\mathbf{r})\right] \cdot \left[\mathbf{r} \times \nabla \left(r^L Y_{LM}\right)\right] d^3\mathbf{r} \tag{A.18}$$

$$= \int_V \nabla \left(r^L Y_{LM}\right) \cdot \left[\mathbf{r} \times \nabla \times \mathbf{j}^*(\mathbf{r})\right] d^3\mathbf{r} \tag{A.19}$$

$$= \int_V \nabla \cdot \left\{\left(r^L Y_{LM}\right) \left[\mathbf{r} \times \nabla \times \mathbf{j}^*(\mathbf{r})\right]\right\} d^3\mathbf{r} - \int_V r^L Y_{LM} \nabla \cdot \left[\mathbf{r} \times \nabla \times \mathbf{j}^*(\mathbf{r})\right] d^3\mathbf{r} \tag{A.20}$$

$$= \oint_S r^L Y_{LM} \left[\mathbf{r} \times \nabla \times \mathbf{j}^*(\mathbf{r})\right] d^2\mathbf{r} - \int_V r^L Y_{LM} \nabla \cdot \left[\mathbf{r} \times \nabla \times \mathbf{j}^*(\mathbf{r})\right] d^3\mathbf{r} \tag{A.21}$$

$$= -\int_V r^L Y_{LM} \left\{\nabla \times \mathbf{j}^*(\mathbf{r}) \cdot (\nabla \times \mathbf{r}) - \mathbf{r} \cdot \left[\nabla \times \nabla \times \mathbf{j}^*(\mathbf{r})\right]\right\} d^3\mathbf{r} \tag{A.22}$$

$$= \int_V r^L Y_{LM} \mathbf{r} \cdot \left\{\nabla \left[\nabla \cdot \mathbf{j}^*(\mathbf{r})\right] - \nabla^2 \mathbf{j}^*(\mathbf{r})\right\} d^3\mathbf{r}. \tag{A.23}$$

By using the continuity Equation [1]

$$i\omega_k \rho(\mathbf{r}) = \nabla \cdot \mathbf{j}(\mathbf{r}), \tag{A.24}$$

Equation (A.23) becomes

$$A^*_{Ele} = \int_V r^L Y_{LM} \mathbf{r} \cdot \left\{(-i)\omega_k \nabla \rho^*(\mathbf{r}) - \nabla^2 \mathbf{j}^*(\mathbf{r})\right\} d^3\mathbf{r} \tag{A.25}$$

$$= -i\omega_k \int_V \nabla \cdot \left[r^L Y_{LM} \rho^*(\mathbf{r}) \mathbf{r}\right] d^3\mathbf{r} + i\omega_k \int_V \rho^*(\mathbf{r}) \nabla \cdot \left(r^L Y_{LM} \mathbf{r}\right) d^3\mathbf{r} \tag{A.26}$$

$$= -i\omega_k \oint_S r^L Y_{LM} \rho^*(\mathbf{r}) \mathbf{r} d^2\mathbf{r} + i\omega_k \int_V \rho^*(\mathbf{r}) \nabla \cdot \left(r^L Y_{LM} \mathbf{r}\right) d^3\mathbf{r} \tag{A.27}$$

$$= i\omega_k (L+1) \int_V r^L Y_{LM} \rho^*(\mathbf{r}) d^3\mathbf{r}. \tag{A.28}$$

For obtaining Eq. (A.14) and Eq. (A.28), we have utilized the following methods and vector calculus identities:

- Integration by part: Eq. (A.17), Eq. (A.20) and the second term of Eq. (A.23).
- Divergence theorem: Eq. (A.18), Eq. (A.21) and Eq. (A.27).
- Identity $\nabla \cdot (\mathbf{a} \times \mathbf{b}) = \mathbf{b} \cdot (\nabla \times \mathbf{a}) - \mathbf{a} \cdot (\nabla \times \mathbf{b})$: Eq. (A.17).
- Identity $\nabla \cdot (\phi \mathbf{a}) = \phi \nabla \cdot \mathbf{a} + \mathbf{a} \cdot \nabla \phi$: Eq. (A.20).

- $\nabla^2\left(r^L Y_{LM}\right)=0$ given by Ref. [1]: the second term of Eq. (A.23).
- Nuclear charge or current is localized: Eqs. (A.18), (A.21) and (A.27).
- Identity $\mathbf{a}\cdot(\mathbf{b}\times\mathbf{c})=\mathbf{c}\cdot(\mathbf{a}\times\mathbf{b})=\mathbf{b}\cdot(\mathbf{c}\times\mathbf{a})$: Eqs. (A.14) and (A.19).
- Curl of the curl $\nabla\times(\nabla\times\mathbf{a})=\nabla(\nabla\cdot\mathbf{a})-\nabla^2\mathbf{a}$: Eq. (A.23).

Then we define the electric multipole moment from Eq. (A.28):

$$\begin{aligned}
\mathbb{Q}_{LM} &= \frac{-A_{Ele}}{i\omega_k(L+1)} && \text{(A.29)}\\
&= \int_V r^L Y^*_{LM}\,\rho(\mathbf{r})d^3\mathbf{r} && \text{(A.30)}\\
&= \langle I_e M_e|\widehat{\mathbb{Q}}_{LM}|I_g M_g\rangle && \text{(A.31)}\\
&= \frac{(-1)^{I_g-M_g}}{\sqrt{2L+1}}\mathrm{C}_{I_g I_e}(L,M;M_g,M_g)\langle I_e|\widehat{\mathbb{Q}}_L|I_g\rangle, && \text{(A.32)}
\end{aligned}$$

and the magnetic multipole moment from Eq. (A.14):

$$\begin{aligned}
\mathbb{M}_{LM} &= -\frac{A_{Mag}}{c(L+1)} && \text{(A.33)}\\
&= \frac{1}{c(L+1)}\int_V [\mathbf{r}\times\mathbf{j}(\mathbf{r})]\cdot\nabla\left(r^L Y^*_{LM}\right)d^3\mathbf{r} && \text{(A.34)}\\
&= \langle I_e M_e|\widehat{\mathbb{M}}_{LM}|I_g M_g\rangle && \text{(A.35)}\\
&= \frac{(-1)^{I_g-M_g}}{\sqrt{2L+1}}\mathrm{C}_{I_g I_e}(L,M;M_g,M_g)\langle I_e|\widehat{\mathbb{M}}_L|I_g\rangle. && \text{(A.36)}
\end{aligned}$$

For obtaining Eqs. (A.32) and (A.36), the Wigner-Eckart theorem is used [2]. The reduced matrix element of the magnetic multipole moment $\langle I_e|\widehat{\mathbb{Q}}_{LM}|I_g\rangle$ and $\langle I_e|\widehat{\mathbb{M}}_{LM}|I_g\rangle$ are now independent of the angular momentum substructure and can be related to the reduced transition probability $\mathbb{B}$ by [2, 3]

$$\begin{aligned}
\mathbb{B}(\varepsilon L) &= \frac{1}{2I_g+1}|\langle I_e\|\widehat{\mathbb{Q}}_L\|I_g\rangle|^2 && \text{(A.37)}\\
\mathbb{B}(\mu L) &= \frac{1}{2I_g+1}|\langle I_e\|\widehat{\mathbb{M}}_L\|I_g\rangle|^2. && \text{(A.38)}
\end{aligned}$$

Finally, the explicit form of Eq. (A.8) is obtained

$$\begin{aligned}
\langle e|\widehat{H}_I(t)|g\rangle = &-i\frac{\sqrt{2\pi}}{\omega_k}e^{-i\Delta_k t}\int_V d^3\mathbf{r}\,\mathbf{j}^*(\mathbf{r})\cdot\sum_{n=-1,1}\left(\mathbf{E}_k\cdot\widehat{\chi}^*_n\right)\\
&\times\left(\widehat{\chi}^*_{-1}\cdot\widehat{\chi}_n\,,\;\widehat{\chi}^*_1\cdot\widehat{\chi}_n\right)\begin{pmatrix}\left(1-\frac{1}{k}\nabla\times\right)\sum_L(i)^L\sqrt{(2L+1)}j_L(kr)\,\mathbf{Y}^{-1}_{LL}\\ \left(-1-\frac{1}{k}\nabla\times\right)\sum_L(i)^L\sqrt{(2L+1)}j_L(kr)\,\mathbf{Y}^{1}_{LL}\end{pmatrix}.\\
&+H.c. \qquad\qquad \text{(A.39)}
\end{aligned}$$

$$= \sqrt{2\pi} e^{-i\Delta_k t} \sum_{n=-1,1} (\mathbf{E}_k \cdot \widehat{\chi}_n^*)$$

$$\times (\widehat{\chi}_{-1}^* \cdot \widehat{\chi}_n \,,\, \widehat{\chi}_1^* \cdot \widehat{\chi}_n) \sum_L \begin{pmatrix} (i)^L \frac{k^{L-1}}{(2L+1)!!} \sqrt{\frac{(2L+1)(L+1)}{L}} \left[\mathbb{M}_{LM}^* + i\mathbb{Q}_{LM}\right] \\ (i)^L \frac{k^{L-1}}{(2L+1)!!} \sqrt{\frac{(2L+1)(L+1)}{L}} \left[-\mathbb{M}_{LM}^* + i\mathbb{Q}_{LM}\right] \end{pmatrix}.$$

$$+ H.c. \tag{A.40}$$

$$= \sqrt{2\pi} e^{-i\Delta_k t} \sum_{n=-1,1} (\mathbf{E}_k \cdot \widehat{\chi}_n^*) \sqrt{2I_g + 1} (-1)^{I_g - M_g}$$

$$\times (\widehat{\chi}_{-1}^* \cdot \widehat{\chi}_n \,,\, \widehat{\chi}_1^* \cdot \widehat{\chi}_n)$$

$$\cdot \sum_L \begin{pmatrix} (i)^L \frac{k^{L-1}}{(2L+1)!!} \sqrt{\frac{L+1}{L}} \left[\sqrt{\mathbb{B}(\mu L)} + i\sqrt{\mathbb{B}(\varepsilon L)}\right] \\ (i)^L \frac{k^{L-1}}{(2L+1)!!} \sqrt{\frac{L+1}{L}} \left[-\sqrt{\mathbb{B}(\mu L)} + i\sqrt{\mathbb{B}(\varepsilon L)}\right] \end{pmatrix}.$$

$$+ H.c.. \tag{A.41}$$

Here $n = 1$ ($n = -1$) for right (left) circularly polarized radiation field.

Derivation of the Maxwell-Bloch Eq. (2.48)

Let's begin with the wave Eq. [4, 5]:

$$\frac{1}{c}\partial_t \mathbf{E} + \partial_y \mathbf{E} = i\frac{2\pi}{\epsilon_0 \lambda} N \mathbf{P} \rho_{eg}, \tag{A.42}$$

where $\mathbf{E}$ the electric field of radiation, ϵ_0 the vacuum permittivity, λ the wave length of resonant radiation, $\mathbf{P}$ the dipole moment of an interacting particle, N the particle number density and ρ the quantum coherence between excited state $|e\rangle$ and ground state $|g\rangle$ of a two-level particle. By performing the following inner product

$$\frac{\mathbf{P}}{\hbar} \cdot \left(\frac{1}{c}\partial_t \mathbf{E} + \partial_y \mathbf{E}\right) = \frac{\mathbf{P}}{\hbar} \cdot \left(i\frac{2\pi}{\epsilon_0 \lambda} N \mathbf{P} \rho_{eg}\right), \tag{A.43}$$

and using the definition of Rabi frequency for the dipole approximation

$$\Omega = \frac{\mathbf{P} \cdot \mathbf{E}}{\hbar}, \tag{A.44}$$

Equation (A.42) becomes

$$\frac{1}{c}\partial_t \Omega + \partial_y \Omega = i\frac{2\pi}{\hbar \epsilon_0 \lambda} N |\mathbf{P}|^2 \rho_{eg}. \tag{A.45}$$

By substituting the so called on-resonance cross section [6]

$$\sigma = \frac{4\pi|\mathbf{P}|^2}{\hbar\epsilon_0\lambda\Gamma}, \tag{A.46}$$

where Γ is the spontaneous decay rate of excited state, into Eq. (A.45), we obtain

$$\frac{1}{c}\partial_t\Omega + \partial_y\Omega = i\eta\rho_{eg}. \tag{A.47}$$

Here $\eta = \frac{\xi\Gamma}{2L}$, $\xi = N\sigma L$ and L is the length of a resonant medium that radiation propagate through.

Steps Between Eqs. (2.84) and (2.87)

$$\begin{aligned}
\Omega_p(t, y) &= \frac{1}{2\pi}e^{-\frac{\Gamma}{2}[\frac{y}{c}+(t-\tau)]}\int_{-\infty}^{\infty} e^{-i[(\frac{\omega}{c}-\frac{\eta}{2\omega})y-\omega(t-\tau)]}d\omega \\
&= \frac{1}{2\pi}e^{-\frac{\Gamma}{2}[\frac{y}{c}+\beta]}\int_{-\infty}^{\infty} e^{i\frac{\eta y}{2\omega}}e^{-i\omega(\frac{y}{c}-\beta)}d\omega \\
&= \frac{1}{2\pi}e^{-\frac{\Gamma}{2}[\frac{y}{c}+\beta]}\int_{-\infty}^{\infty} e^{i\frac{q}{\omega}}e^{-i\omega z}d\omega \\
&= \frac{1}{2\pi}e^{-\frac{\Gamma}{2}[\frac{y}{c}+\beta]}\sum_{n=0}^{\infty}\frac{(iq)^n}{n!}\int_{-\infty}^{\infty}\frac{1}{\omega^n}e^{-i\omega z}d\omega \qquad \text{(A.48)} \\
&= \frac{1}{\sqrt{2\pi}}e^{-\frac{\Gamma}{2}[\frac{y}{c}+\beta]}\sum_{n=0}^{\infty}\frac{(iq)^n}{n!}\left[-i\sqrt{\frac{\pi}{2}}\frac{(-iz)^{n-1}}{(n-1)!}sgn(z)\right] \qquad \text{(A.49)} \\
&= \frac{1}{2}e^{-\frac{\Gamma}{2}[\frac{y}{c}+\beta]}sgn(z)\sum_{n=0}^{\infty}\frac{q^n z^{n-1}}{(n-1)!n!} \\
&= \frac{1}{2}e^{-\frac{\Gamma}{2}[\frac{y}{c}+\beta]}sgn(z)q\sum_{n=0}^{\infty}\frac{(-qz')^{n-1}}{(n-1)!n!} \\
&= e^{-\frac{\Gamma}{2}[\frac{y}{c}+\beta]}\delta(z) + \frac{1}{2}e^{-\frac{\Gamma}{2}[\frac{y}{c}+\beta]}sgn(z)q\sum_{n=1}^{\infty}\frac{(-qz')^{n-1}}{(n-1)!n!} \qquad \text{(A.50)} \\
&= \delta(z)e^{-\frac{\Gamma}{2}[\frac{y}{c}+\beta]} - q\frac{J_1\left(2\sqrt{qz'}\right)}{2\sqrt{qz'}}e^{-\frac{\Gamma}{2}[\frac{y}{c}+\beta]}. \qquad \text{(A.51)}
\end{aligned}$$

Appendix B
Effective Laser Pulse

In Sect. 3.2.1, the concept of the effective intensity [2] has been introduced, whereas the effective temporal shapes of the incident pulses are not addressed. Here we discuss this interesting question about the effective pulse shape which may be observed by a nucleus. Considering a fully coherent XFEL Gaussian pulse $Ie^{-(\Gamma t)^2}$, where I and Γ are the peak intensity and the bandwidth, respectively. The Fourier cosine transform of the XFEL pulse is

$$\frac{I}{\sqrt{2\pi}}\int_{-\infty}^{\infty} e^{-(\Gamma t)^2}\cos(\omega t)\,dt = \frac{I}{\sqrt{2}\Gamma}e^{-\left(\frac{\omega}{2\Gamma}\right)^2}. \tag{B.1}$$

For obtaining the effective laser pulse, we calculate the inverse Fourier cosine transform of Eq. (B.1) for a narrow frequency domain $(-f\Gamma, f\Gamma)$ and expand the result for $f \ll 1$

$$\begin{aligned} &\frac{I}{2\Gamma\sqrt{\pi}}\int_{-f\Gamma}^{f\Gamma} e^{-\left(\frac{\omega}{2\Gamma}\right)^2}\cos(\omega t)\,d\omega \\ &= \frac{I}{2}e^{-(\Gamma t)^2}\left[\operatorname{erf}\left(\frac{f}{2} - i\Gamma t\right) + \operatorname{erf}\left(\frac{f}{2} + i\Gamma t\right)\right] \qquad \text{(B.2)} \\ &= I\frac{f}{\sqrt{\pi}}\left\{1 - \frac{f^2}{12}\left[1 + 2(\Gamma t)^2\right] + \frac{f^4}{480}\left[3 + 12(\Gamma t)^2 + 4(\Gamma t)^4\right]\ldots\right\}, \qquad \text{(B.3)} \end{aligned}$$

where f is the ratio of a nuclear linewidth to the laser bandwidth, and erf(z) is the error function [7]. Equation (B.3) gives the effective laser pulse that may interact with a nucleus as its considered transition linewidth is narrower than the bandwidth of an incident XFEL pulse. For an extremely narrow nuclear linewidth ($f \ll 1$), the incident pulsed laser may effectively become a CW laser as showed by the first term of Eq. (B.3). If the considered f is not so small, the other terms will modify the effective pulse shape which is still longer than that of the incident one. Remarkably, this effect is "spooky" because Eq. (B.3) reveals that a nucleus may already interact with a very short laser pulse before the pulse arrives at the nucleus. This interesting

W.-T. Liao, *Coherent Control of Nuclei and X-Rays*, Springer Theses,
DOI: 10.1007/978-3-319-02120-1,

effect is due to the wave nature of the radiation, and may be tested in ultrafast optics of nowadays. Also, longer effective XFEL pulses will reduce the required laser intensity for NCPT and lower the threshold for experiments.

References

1. J.M. Blatt, V.F. Weisskopf, *Theoretical Nuclear Physics* (Courier Dover Publications, New York, 1991)
2. A. Pálffy, J. Evers, C.H. Keitel, Electric-dipole-forbidden nuclear transitions driven by super-intense laser fields. Phys. Rev. C **77**, 044602 (2008)
3. P. Ring, P. Schuck, *The Nuclear Many-Body Problem* (Springer Verlag, New York, 1980)
4. M.D. Crisp, Propagation of small-area pulses of coherent light through a resonant medium. Phys. Rev. A **1**, 1604 (1970)
5. P.-C. Guan, I.A. Yu, Simplification of the electromagnetically induced transparency system with degenerate zeeman states. Phys. Rev. A **76**, 033817 (2007)
6. D. Steck, Rubidium 87 d line data, http://steck.us/alkalidata/ (2001)
7. G. Arfken, H. Weber, F. Harris, *Mathematical Methods For Physicists International Student Edition* (Academic Press, New York, 2005)